AF767225

Hydrology - The Science of Water

Edited by Muhammad Salik Javaid

Published in London, United Kingdom

IntechOpen

Supporting open minds since 2005

Hydrology - The Science of Water
http://dx.doi.org/10.5772/intechopen.78508
Edited by Muhammad Salik Javaid

Contributors
Cheikh Faye, Amin Shaban, Nawa Raj Pradhan, Charles Downer, Sergei Marchenko, Mikhail Burakov, Milan Gombos, Branislav Kandra, Andrej Tall, Dana Pavelková, Vanessa J. Banks, Barbara Palumbo-Roe, Catherine Russell, Muhammad Salik Javaid

© The Editor(s) and the Author(s) 2019
The rights of the editor(s) and the author(s) have been asserted in accordance with the Copyright, Designs and Patents Act 1988. All rights to the book as a whole are reserved by INTECHOPEN LIMITED. The book as a whole (compilation) cannot be reproduced, distributed or used for commercial or non-commercial purposes without INTECHOPEN LIMITED's written permission. Enquiries concerning the use of the book should be directed to INTECHOPEN LIMITED rights and permissions department (permissions@intechopen.com).
Violations are liable to prosecution under the governing Copyright Law.

Individual chapters of this publication are distributed under the terms of the Creative Commons Attribution 3.0 Unported License which permits commercial use, distribution and reproduction of the individual chapters, provided the original author(s) and source publication are appropriately acknowledged. If so indicated, certain images may not be included under the Creative Commons license. In such cases users will need to obtain permission from the license holder to reproduce the material. More details and guidelines concerning content reuse and adaptation can be found at http://www.intechopen.com/copyright-policy.html.

Notice
Statements and opinions expressed in the chapters are these of the individual contributors and not necessarily those of the editors or publisher. No responsibility is accepted for the accuracy of information contained in the published chapters. The publisher assumes no responsibility for any damage or injury to persons or property arising out of the use of any materials, instructions, methods or ideas contained in the book.

First published in London, United Kingdom, 2019 by IntechOpen
IntechOpen is the global imprint of INTECHOPEN LIMITED, registered in England and Wales, registration number: 11086078, The Shard, 25th floor, 32 London Bridge Street
London, SE19SG – United Kingdom
Printed in Croatia

British Library Cataloguing-in-Publication Data
A catalogue record for this book is available from the British Library

Additional hard and PDF copies can be obtained from orders@intechopen.com

Hydrology - The Science of Water
Edited by Muhammad Salik Javaid
p. cm.
Print ISBN 978-1-83962-646-3
Online ISBN 978-1-83880-324-7
eBook (PDF) ISBN 978-1-83881-852-4

For EU product safety concerns:
IN TECH d.o.o., Prolaz Marije Krucifikse Kozulić 3, 51000 Rijeka, Croatia, info@intechopen.com or visit our website at intechopen.com.

We are IntechOpen, the world's leading publisher of Open Access books
Built by scientists, for scientists

4,300+
Open access books available

116,000+
International authors and editors

125M+
Downloads

151
Countries delivered to

Our authors are among the
Top 1%
most cited scientists

12.2%
Contributors from top 500 universities

BOOK CITATION INDEX
WEB OF SCIENCE™

Selection of our books indexed in the Book Citation Index
in Web of Science™ Core Collection (BKCI)

Interested in publishing with us?
Contact book.department@intechopen.com

Numbers displayed above are based on latest data collected.
For more information visit www.intechopen.com

Meet the editor

Dr. Muhammad Salik Javaid has a long meritorious career interacting with "water" as an engineering student, a practicing professional engineer, a researcher, and as a university teaching faculty member. For over three decades he has worked in Pakistan and abroad for the Corps of Engineers, the Earthquake Reconstruction and Rehabilitation Authority, the Frontier Works Organization, (and the Design and Consultancy Department. His undergraduate engineering education is from the Military College of Engineering. He is a graduate of Georgia Tech, Atlanta, USA, where he earned his Master's and Doctoral degrees in Civil Engineering. He has been on the faculty of the National University of Sciences and Technology, the University of Lahore, and Abasyn University teaching undergraduate- and graduate-level courses related to water.

Contents

Preface

The importance of the science of water, commonly called "the hydrologic science," is increasing day by day as the resources of water are depleting and becoming scarcer. The depletion problem is not that the amount of total water available on this planet is reducing with time, but that the amount of sweet water, potable water, arable water, usable water, and water suitable for human usage and consumption is depleting every day in contrast to the rapid population growth on this planet. This is happening because the resource is being polluted by humankind itself, rendered unsuitable due to negligence and mismanagement, or its quantity is fast diminishing due to exhaustive extraction and consumption.

With this state of affairs, in recent times, knowledge of the science of water has gained an importance many times its original scale; and with that, its acquisition, expansion, research, advancement, and dissemination have become equally or more important. With so many dimensions of hydrology, any contributory work on the science of water, now available for exploration, research, and technological advancement, is more than welcome. This book will play its part in furthering the knowledge of the science of water and will prove useful reading for various cross-sections of academia in training institutions, researchers in laboratories, and engineers working in the field.

This voluminous book is the outcome of the consistent work of authors, researchers, engineers, and scientists from various continents across the globe. It consists of seven chapters covering various dimensions of hydrology. The first chapter jointly authored by myself and Ms. Laila Khalid, "Hydrologic science—where we stand today?," presents the journey of progress of hydrology through the ages, culminating in the present-day state of knowledge, research, and application of hydrologic science. The next chapter titled "The hyporheic zone" by Vanessa J. Banks introduces the key concepts of the hyporheic zone with an emphasis on the importance of understanding streambed sediments and their architecture to assess hydraulic functioning and modeling of such a zone. The third chapter by Mikhail Burakov discusses the accuracy of hydrogeological calculations and forecasts, and presents a method to make assessments of flow parameters. The fourth chapter authored by Faye Cheikh tries to assess the impacts of climate change on the water resources of the Gambian River Basin. In Chapter 5 "Analysis of non-rainfall periods and their impacts on soil water regime," Milan Gombos investigates significant non-rainfall periods, their periodicity, and statistical characteristics in the region of the eastern Slovak lowlands. The sixth chapter written by Amin Shaban, "Striking challenges on water resources of Lebanon," investigates the existing physical and anthropogenic challenges faced by water resources in Lebanon, even though there is no visible shortage of water in the region. The seventh chapter by Nawa Raj Pradhan, "Process modeling of soil thermal and hydrological dynamics," explains the simulation of soil thermal state effects on hydrological response, soil thermal regime, frozen soils, and permafrost simulation capability of a model developed at a permafrost laboratory.

The entire concept of IntechOpen Access is simply laudable. Providing a platform for writers, researchers, scholars, scientists, engineers, users, and all stakeholders of "water" to give them an opportunity to document their valuable knowledge is a great scheme in itself. Also, providing an open and free access of this knowledge to the whole world is the greatest service possible for humanity. In the offices of IntechOpen the contributions of author service managers Ms. Manuela Gabrić and Mr. Josip Knapić need to be recognized because without their relentless reminders of the timelines, their cooperation, and their assistance this book would not have been completed within the specified time limits.

The contributions of my colleague Dr. Fariha Masood Siddiqui are also recognized as it was her knowledge and comments in the field of water quality that helped me complete the editing job in a befitting manner encompassing all needed aspects.

I dedicate this book to my grandson Ibrahim Waqar and all kids of his age. It is for his generation and all generations that come after him that we would want to leave this already scarce, life-sustaining resource called water in a fit, suitable, usable, and sustainable form. The dangers of global warming, drastic climate change, ozone layer depletion, and water and food scarcity are looming over the coming generations and need to be challenged and remedied, right now! May the future generations live in an environment much better than ours.

Dr. Muhammad Salik Javaid
Abasyn University,
Islamabad, Pakistan

Chapter 1

Introductory Chapter: Hydrologic Science - Where We Stand Today?

Muhammad Salik Javaid and Laila Khalid

1. A journey through ages

Much has been written in the scientific literature about this diverse field of hydrologic science. However, this chapter is aimed at introducing modern hydrology, highlighting its evolution from ancient times to the recent era of rationalization, theorization, and computerization. Though hydrology is a main branch of science, it has exhibited very slow but steady developmental strides as compared to some other branches of science showing quantum leaps. Discussing the historical development of hydrology, it can be divided into [1] six successive periods of time: (a) ancient hydrology (1000 BC–AD 500), (b) medieval water technology (500–1500), (c) renaissance concepts and measurements (1500–1700), (d) the emergence of water science (1700–1800), (e) the era of empirical hydrology (1800–1930), and (f) rationalization, theorization, and computerization (1930–2000).

Hydrology is the science of water came into being the day the first winds blew across the meadows, the day first cloud was formed in the sky, the day the first drop of rain fell on the surface of this earth, the day some sage tried to collect rainwater and diverted it to his farmlands, and the day mankind got an inkling of the immense value of water for the survival of life. Now once we look back, we see that starting from that very first day, the hydrology has come a long way, travelling through history and adopting many faces of knowledge. In the thematic framework, hydrology now has evolved to become a cross-cutting theme for civil engineering, agricultural engineering, earth science, environmental science, climatology, geography, geology, and watershed management. Hydrologic phenomena have proved to be the most intriguing among elements of our environment, that is, sun, soil, air, and water.

During the eighteenth century, there was a huge advancement in the field of hydrology, when new hydraulic principles and experiments such as Bernoulli's equation and Chezy's formula, and improved instruments such as current meter and tipping bucket rain gauge, were developed. John Dalton derived the principle for evaporation using lysimeter, and Mulvaney proposed the rational method for determining peak flood flow. Darcy derived his equation and developed his law of porous media flow, while Chamier applied rational method to the design of culverts, and Kuichling applied rational method for the estimation of storm water flows in urban areas.

In the nineteenth century, to meet the estimation and design demands of hydraulic engineering projects, empirical approach was used. However, at the start of the twentieth century, the quantitative hydrology was still undeveloped. The year from 1930 to 2000 is the period in which modern hydrology evolved through stages and developed as a geoscience. In this period, Horton developed infiltration theory, and Richard developed the equation for unsaturated flow. In the early 1930s, the main focus was rationalization of empirical results which were derived earlier and used in practice. In the middle of the period, theoretical basis replaced empiricism,

IntechOpen

and government agencies started their own hydrologic research programs. During this time, the concept of hydrograph and more specifically the concept of unit hydrograph were developed. This was the first approach during which precipitation and runoff were used to check the rapid response of drainage basin.

In the period following that hydrology has been approached with a more theoretical basis than in the past, numerical models and digital computers were used to solve the hydraulic engineering projects.

2. Role of hydrological modeling

In the hydrologic system, a lot of changes have occurred in the recent past due to climate change, rapid urbanization, and industrialization. Climate change can significantly affect water resources by changing the hydrological cycle. The change in temperature and precipitation has direct effect on evapotranspiration, which directly affects the runoff generation. Some important indicators of climate change are droughts, floods, shifting seasons, melting of polar ice caps, changes in the pattern of oceanic currents, and lowering of water table [2]. In order to overcome these challenges, various hydrological models have been developed to investigate, understand, and explore solutions for sustainable water management. Hydrological models are basic and theoretical representations of the hydrologic cycle. They are mainly used for the prediction and understanding of hydrological processes. Hydrological models may be categorized as:

a. models based on data collection

b. models based on process description.

Models based on data are sometimes considered like black box systems. Shahid et al. [3] have illustrated the use of double mass curve in rainfall-discharge-sedimentation studies. Mathematical and statistical concepts are used in these models to link a certain input to the model output. Transfer function, system identification, regression, and neural network are commonly used techniques. Linear models are the simplest models. These models are known as stochastic hydrology models.

Models based on process description are more complicated than the stochastic hydrological models. Physical processes that occurred in the real world are represented through these models, such as representation of subsurface flow, channel flow, evapotranspiration, and surface runoff. Models based on process are known as deterministic hydrology models.

3. GIS applications in hydrological science

GIS is considered as one of the most important new technologies with the potential to revolutionize many aspects of society through increased ability to decisions and solve problems. GIS provides a digital representation of watershed characteristics used in hydrologic modeling. For very large areas, the main advantage of GIS is its ability to integrate, analyze, and manage large volume of data [4]. The applications of GIS in hydrology are mainly useful for watershed analysis such as surface and groundwater modeling, regional groundwater modeling, and water quality analysis. Hydrological models require huge volume of input data like soil cover data, topography, weather, and geologic data. The ability of GIS to integrate data from multiple sources such as boreholes and wells, subsurface isopach (or structure contour) maps, and surface geology maps also allows data to be used simultaneously to develop a groundwater model. Water quality issues associated with regional changes in land use, such as urbanization and large-scale agriculture, can also be analyzed effectively using GIS. On a local scale, GIS can be useful for landfill siting and the selection of groundwater development and artificial recharge sites.

Remote sensing and GIS are integral to each other. Remote sensing is a technique to observe the earth's surface or the atmosphere from outer space using satellites (space-born) or from the air using aircrafts (airborne). Weather radar is used to determine speed and direction of wind, meteorology, climatology, and the study of ionosphere. Altimeters make use of remote sensing for measuring ocean currents, wavelength, their direction, and water level. Light detection and ranging (LIDAR) is used for vegetation sensing and for measurement of concentration of various chemicals in atmosphere and heights of different objects [5]. A number of major improvements have been made to existing hydrologic models, such as HEC-HMS, HEC-RAS, and EPA SWMM.

4. Role of artificial intelligence in hydrological science

During the last decades, mostly physical models were used for almost all hydrological investigations. These models give accurate results if detailed and accurate data are available; but in developing countries, accurate data are not available due to the financial and technical constraints, which results in model uncertainties and unsatisfactory performance. To overcome these shortcomings, artificial intelligence (AI)-based techniques have been recently used as alternative tools to traditional physical hydrological models [6]. Nowadays, artificial intelligence has been used in numerous hydrologic modeling such as rainfall prediction, rainfall-runoff modeling, evaporation estimation, flood forecasting, river stage prediction, sediment load modeling, and water quality simulation. Many techniques can be categorized under this newly available approach such as artificial neural networks (ANNs) and support vector machine (SVM). These techniques have become very popular and efficient for modeling complicated hydrological processes using relatively less cost, effort, and data. These approaches have proven to be highly efficient in solving the problems when the rules to solve such problems are difficult to be expressed which is the case in most hydrological processes. Therefore, AI techniques have become attractive alternatives to traditional physical modeling techniques for many hydrological applications. The position at the end of the twentieth century was that hydrology had developed to the state that it could be considered as a science whose analytical findings could be applied with profit to the solution of social problems.

5. Where we stand today?

Nevertheless, the most important lesson of this chapter is that observational hydrology is as old as human civilization. Second, our theoretical debts are to physicists, microbiologists, chemists, and mathematicians. Hydrologists used their hard-won observations to build complex water projects, but they did not create the important theories. Modern hydrology now emphasizes computer models over real example and illustrates important concepts with cartoons, while abundant real-world data are at our fingertips. In our view, future advancements of the hydrologic sciences will depend on a strong observational foundation consistent with its roots. This foundation enabled the large-scale water projects of ancient civilizations and is equally essential today.

The application of artificial intelligence, artificial neural networks, support vector machine, fuzzy logic, wavelets, entropy theory, network theory copula theory, chaos theory, and catastrophe theory are mostly in practice during the last two decades. With advances in data capturing, analysis, and transfer, it seems that the future of hydrology will be even brighter, with new tools at its disposal. Of course, computers have increased our ability to design systems, process data, and envision complex processes [7]. Hydrologic models will become so user-friendly that little hydrologic knowledge will be needed to operate them, just like one does not need to be an automobile engineer to drive a car or an electrical engineer to operate an electrical system. However, the models are simple or complicated, and these will be associated with a statement of uncertainty. Modern hydrology will play a great role to overcome the mega challenges of this century such as environmental security, water security, energy security, food security, water-food-energy nexus, health security, and sustainable development.

The water laws, water legislation, water apportionment, and water rights of various users and stakeholders leading to water conflicts, water wars, and conflict resolution mechanisms and strategies are the oldest as well as the latest trends in the hydrological sciences. Studies in transboundary aquifers and water bodies situated across the state and nation lines are the focus of research for many regional, national, and international; political and financial; bodies and forums. The physical, chemical, and medical health of such water bodies, their maintenance, and upkeep are the current issues of mutual cooperation as well as mutual conflicts. The spine of such mutual cooperation and mutual conflict resolution is the availability of database of water resources and their historical usages, giving rise to the importance of such neutral data repositories at national, regional, and international levels.

Education and awareness about the conservation and maintenance of water resources for all communities and stakeholders is another very important aspect of hydrological sciences. The advancement in today's modern technologies in the fields of electronics, computers, nanomaterials, medical, and energy, etc. has proven to be very fascinating and eye catching for the young brains. Hydrological sciences and the like have somehow gone on the backburners, getting not much attention. The old concept of "seven seas filled with water" needs to be reviewed by all with a concerted awareness and educational campaigns for saving and preserving this resource for the generations to come.

6. Conclusion

Hydrology, the science of water, is a multi-faceted science with branches like engineering hydrology, groundwater hydrology, surface hydrology, geohydrology, ecohydrology, hydrometeorology, hydroinformatics, statistical hydrology, and stochastic hydrology. Although the knowledge about science of water is available in abundance, which is spread all over in these disciplines, there is a need to harness the explorations of new vistas, latest developments, efficient frameworks, and cutting-edge technologies and collect these in one place, one volume, and one book so that all the experts, stakeholders, academia, and field professionals have a fair and free access.

Author details

Muhammad Salik Javaid[1]* and Laila Khalid[2]

1 Civil Engineering Department, Abasyn University, Islamabad, Pakistan

2 Capital University of Science and Technology, Islamabad, Pakistan

*Address all correspondence to: msalikj@abasynisb.edu.pk

IntechOpen

© 2019 The Author(s). Licensee IntechOpen. This chapter is distributed under the terms of the Creative Commons Attribution License (http://creativecommons.org/licenses/by/3.0), which permits unrestricted use, distribution, and reproduction in any medium, provided the original work is properly cited.

References

[1] Dooge JC. Background to modern hydrology. The basis of civilization– Water science. In: Proceeding of UNESCO/IAHS Symposium; December 2003. Rome: IAHS Publication 286; 2004. pp. 3-12

[2] Vörösmarty CJ, Green P, Salisbury J, Lammers RB. Global water resources: Vulnerability from climate change and population growth. Science. 2000;**289**(5477):284-288

[3] Shahid M, Gabriel HF, Nabi A, Javaid MS. Assessment of effect of land use change on hydrological response and sediment yield for catchment area of Simly Lake, Pakistan. In: Proceedings of 1st International Conference on Emerging Trends in Engineering, Management and Sciences (ICETEMS-2014); 29-30 December 2014; Islamabad, Pakistan. 2014

[4] De Vantier BA, Feldman AD. Review of GIS applications in hydrologic modeling. Journal of Water Resources Planning and Management. 1993;**119**(2):246-261

[5] Tripathi MP, Panda RK, Pradhan S, Sudhakar S. Runoff modelling of a small watershed using satellite data and GIS. Journal of the Indian Society of Remote Sensing. 2002;**30**(1-2):39

[6] Alagha JS, Said MAM, Mogheir Y. Artificial intelligence based modelling of hydrological processes. In: 4th International Engineering Conference– Towards Engineering of 21st Century. 2012

[7] Singh VP. Hydrologic modeling: progress and future directions. Geoscience Letters. 2018;**5**(1):15

Chapter 2

The Hyporheic Zone

Vanessa J. Banks, Barbara Palumbo-Roe and Catherine E. Russell

Abstract

This chapter introduces the key concepts of the hyporheic zone. It considers the research context in terms of the Water Framework Directive and the breadth of literature associated with the hyporheic zone. The interplay between hydrological, chemical and biological processes is explained, and a range of different approaches to field sampling and monitoring are described. A framework for considering the factors contributing to the conceptualisation of the hyporheic zone is presented, with an emphasis on the importance of understanding streambed sediments and their architecture to assess hydraulic functioning and modelling of the hyporheic zone. The hyporheic zone in karst catchments is also given specific consideration. Returning to the theme of linked hydrological, biological and chemical processes, the results of two case studies demonstrate the value of integrating hydrological measurements with geochemistry in order to elucidate hyporheic zone functioning.

Keywords: surface water groundwater interaction, hyporheic zone, monitoring

1. Introduction

1.1 Aims and objectives

The aim of this chapter is to consider the hydrology of the hyporheic zone in the context of the hydrological cycle. The chapter starts, the chapter starts with a description of the definitions of, and reasons for studying, the hyporheic zone, then goes on to consider sampling and measurement techniques followed by a consideration of the assessment and applications of this data and understanding. The results of two case studies provide examples of research approaches and technique. The case studies show how integrating hydrological measurements with geochemistry assists in the elucidation of hyporheic zone functioning.

1.2 Definitions and significance

The hyporheic zone is the term given to the subsurface interface between surface and groundwater bodies. It is most commonly considered in the context of streams (or rivers) interfacing with groundwater. Groundwater in this context is the water that fills the spaces between soil particles and fractured rock that comprises the saturated ground that extends beneath the water table and the overlying unsaturated zone. Cardenas [1] suggests that the hyporheic zone is synonymous with the transient storage zone of earlier literature, i.e. including bank storage within the riparian zone (broadly defined as the interface between land and a river or stream), e.g. Bencala and Walters [2]. As it is a zone of flux between ground- and surface water, there are hydrological, ecological and hydrogeological responses that characterise the hyporheic zone. Reflecting the range of perspectives, there are differing definitions of the hyporheic zone. Hydrologically it is conceptualised as the proportion of flow that occurs in permeable streambed deposits upon which channel flow occurs. This can also be seen as the component of flow that cannot be measured using conventional flow monitoring techniques. The sediments in this zone have an important role in influencing the distribution of permeability in the hyporheic zone attributable to grain size distribution, source rock and architecture, as related to topography, river dynamics and climate [3] and modified by biological and chemical processes. Orghidan [4] recognised the ecological significance of the hyporheic zone with the introduction of the term "hyporheic corridor concept". One difficulty for the ecologist is defining the thickness of the hyporheic zone. For example, they can delineate it by the occurrence of hyporheobiont life stages or by the extent of riverine animals [5]. The definition favoured by Brunke and Gonser [6] is that the hyporheic zone is distinguished from groundwater and stream water by demonstrating characteristics of both, with different gradients to each. Clearly the hyporheic zone is dynamic as a consequence of changing hydraulic conditions and seasonality, and this is recognised by ecologists in the term "dynamic ecotone" [5]. The hydrogeologist's view of the hyporheic zone is as part of the groundwater system, because it comprises subsurface water within the saturated zone.

As a concept, the hyporheic zone is important in addressing integrated catchment modelling and management. In Europe the Water Framework Directive (2000) provides the context for an increased research interest in the hyporheic zone [7], because it promotes the management of groundwater bodies and surface water bodies in an integrated way, requiring that (hydraulic) pathways between the two are understood. Analysis of the connectivity of surface and groundwater in conjunction with other protected areas such as designated wetland is a specific requirement of the River Basin Catchment Plan [5]. The ability to assess mass flux across the groundwater-surface water interface, predict attenuation processes in this zone, link hyporheic and benthic chemical conditions and ecological health and develop reliable and transferable conceptual models of flow and attenuation defines the requirements of conceptual understanding.

Soil properties impose a strong influence on the dynamics of the hyporheic zone in terms of transient storage and retention. The transient storage capacity of the hyporheic zone can be important in accounting for apparent losses or gains in water balance calculations, e.g. Lapworth et al. [8], which may inform resource evaluation studies. Improved understanding of the spatial components of the hyporheic zone offers significant potential in terms of understanding the process of flood migration along the length of the stream in the context of catchment scale flood modelling and management. Hydrologically, the hyporheic zone is an important component of some poorly understood karst systems (e.g. turloughs or estavelles), wetlands and lake environments. More recently, in the context of urban environments, the concept of the hyporheic zone has been extended to include the impacts of leaking pipes contributing water of different chemistry to the zone of transient storage or service ducts providing preferential pathways, perched water tables and altered flow and groundwater conditions, e.g. Bricker et al. [9].

Within the hyporheic zone, the geology, hydrology, hydrochemistry and biology exhibit feedbacks and dependencies. Consequently, hydrological understanding is important to the aspects of hyporheic zone research that embrace the ecological and chemical benefits (ecosystem services) of the zone. Broadly, hyporheic faunal communities vary with the environmental conditions, including hydrology, climate, geology, sedimentary architecture, land use and chemical conditions (natural and anthropogenic). The influence of hydrological flux in the hyporheic zone is particularly important in defining nutrient (carbon) distribution and its upward and downward movement and consequently the distribution of the ecotones that impose structure on the hyporheic communities (hyporheos, including benthic, epigean and phreatic species) and their distribution in the sediment [10, 11]. For example, during periods of environmental stress, typically marked by drought or flood, the hyporheic zone provides a place of refuge for some stream-dwelling species. Other species permanently occupy the hyporheic zone niche. In some streams, the hyporheic zone species extend in excess of 100 m beneath the streambed, e.g. [12]; elsewhere they occur at relatively shallow depths. There have been a number of associated shifts in understanding relating to the discovery of hyporheic invertebrates and the concept of the hyporheos as an indicator of ecological health. Furthermore, the broad range of biological species, including microbial fauna, has an important role in contaminant attenuation, and this defines the zone as an ecosystem service with a potential to mitigate contaminants. Whilst there is an extensive literature associated with this topic, it lies beyond the remit of this chapter.

In conjunction with a range of biological processes, the geochemistry of the hyporheic zone provides a valuable natural system for the remediation of a range of contaminants. For example, this zone is particularly important in hosting denitrification processes. There are a number of factors that contribute to the geochemistry of the hyporheic zone, including bedrock geology, superficial geology, water residence time, oxygen concentrations, the degree of mixing of ground- and surface water, pH conditions and breakdown of contaminants, including plastic and the organic content. Additionally, bedrock textures may be important for the growth of specific precipitates. Its attenuation capacity varies with its thickness and permeability. This potential has been recognised in the context of the remediation of the legacy of abandoned mines in the UK [13].

2. Research methods: hydrological measurements and sampling in the hyporheic zone

As with any other system, sampling and monitoring of the hyporheic zone requires a strategy and plan [14], the formulation of which requires a clear understanding of the reason for monitoring and what it aims to achieve. This is particularly relevant in the context of the hyporheic zone where, if it is required, it may be possible, with careful planning, to optimise sampling to derive hydrological, ecological and hydrogeochemical data in conjunction with each other. The second step in the development of a sampling strategy is the completion of a desk-based study of the area of interest. Ideally, this should be undertaken at the catchment scale to understand the broader hydrological context with subsequent more detailed studies at the sub-catchment, reach and project scale. As well as considering the spatial scale of interest, decisions will have to be made regarding the temporal aspects of data collection: how frequently will data be collected and how long will the monitoring continue in order to characterise the flow regime, chemical and biological context? Subsequent decisions will relate to how to undertake the monitoring or sampling and whether it should comprise point methods, averaging methods or distributed methods to provide insight into spatial or temporal variation [14]. Key factors influencing the frequency, duration and type of monitoring include the funding that is available, site access and health and safety considerations as well as the scientific factors, such as the nature of geology proposed for sampling. Streambeds with rock or coarse sediments in their base are inherently more challenging than finer sediment-bedded streams. A selection of potential sampling methods is detailed in **Table 1**. In selecting appropriate sampling and monitoring

Method	Description	Hydrological	Ecological	Hydrogeochemical
Seepage meter [15]	Measures exchange of water across the sediment–water interface	Y		
Wells and sampling pits	Can be used for sediment characterisation as well as constructing access for water sampling. Can also be used for hydraulic testing (e.g. falling or rising head tests; slug tests). The sampling zone of a well is dictated by the positioning of the sample points, e.g. well screen	Y		Y
Mini drive-point piezometers [16]	Used to enhance understanding of head and hydrochemistry (see case studies)	Y		Y
Natural tracers including temperature and electrolytic conductivity	Measurement of temperature as a tracer of flow paths. Care needed to ensure that readings are not affected by sampling materials, if measured in a piezometer or multilevel sampler	Y		
Physicochemical parameters as a proxy for hydrochemical zoning	pH, temperature, electrolytic conductivity, dissolved oxygen and redox probes			Y
Diffusive equilibrium in thin films (DET), Byrne et al. [17], or diffuse gradient in thin films [18]	In situ passive sampling of porewater by diffusive equilibrium in thin films. Stainless steel cover containing DET gel			Y
Net sampling	Can be lowered into wells or natural water bodies		Y	
Kick sampling [19]	Standard method in qualitative studies of macroinvertebrates		Y	
Pump sampling [20]	Another method for qualitative studies of macroinvertebrates		Y	
Automatic sampling	Automatic samplers can be programmed to take periodic samples, e.g. to sample through a weather event such as a storm		Y	Y
Kubiena tin or similar	Used for undisturbed sampling of sediment for resin moisture replacement and optical examination of hydrogeological properties	Y		
Geotechnical soil property tests	For geotechnical characterisation of the alluvium, might include density, grading, porosity, field capacity, hydraulic conductivity, moisture content and electrical resistivity	Y		

Method	Description	Hydrological	Ecological	Hydrogeochemical
Geophysical testing [21]	Field-based electrical resistivity can be used in monitoring contaminant plumes or tracers such as salt, as well as monitoring changes in moisture content and the position of the water table	Y		Y

Table 1.
A range of potential sampling/monitoring methods.

techniques, construction materials and their potential to impact results should be considered; for example, whilst robust, the use of galvanised steel products can give rise to artificially high concentrations of zinc.

Catchment scale analysis of the topography, land use, geology, hydrogeology and water features, as well as any hydrological data, will provide an evidence base for assessing potential zones of groundwater-surface water interaction and possible access points to the stream. Typically, hydrological studies focus on the flux and determination of the gradients between the surface and groundwater; therefore, they will largely focus on head, seepage, changes in stream discharge, permeability and hydraulic conductivity. This will require measurements in the subsurface, in the hyporheic zone and in the stream. Wherever possible measurement techniques should be selected to provide complimentary data and optimise confidence in the results.

Traditionally, flow regimes within a single river system were measured using flow-gauging techniques, typically using impeller or electromagnetic current meters [22]. However, whilst these techniques are valuable in identifying zones of potential loss or gain in discharge for further investigation, they preclude measurement of flow in the hyporheic zone; therefore, different or additional techniques are required for monitoring the hyporheic zone. Three-dimensional temperature has been found to be a valuable parameter for characterising the hyporheic zone [23, 24]. This is because the longer the groundwater is in contact with bedrock, the more it will equilibrate with the bedrock temperature, whereas surface water will tend to equilibrate with atmospheric conditions. The rates of temperature exchange are affected by the depth of the river, the thickness and permeability of the river sediment, the characteristics of the upper part of the bedrock and the specific heat capacities of the bedrock and sediments. These effects vary both seasonally and in response to hydrological events such as flooding or, for example, releases of dam water [24, 25].

Tracing experiments provide the advection and dispersion characteristics of the transient storage zone that are required for modelling [26]. Conservative tracers, such as sodium chloride, sodium bromide and rhodamine WT are well-established techniques in the analysis of hydrological characteristics of streams [27, 28]. Others have included natural tracers such as radon [29]. Additionally, there are a number of emerging, anthropogenically introduced conservative tracers; for example, Möller et al. [30] used gadolinium. Other contaminants can be utilised as tracers to demonstrate the extent of the hyporheic zone or the extent of change in the hyporheic zone, e.g. Ciszewski and Aleksander-Kwaterczak [31] used the concentrations of zinc and cadmium in the sediment and waters of Baila Przemsza River in southern Poland to define the extent of mining-induced alteration of the hyporheic zone. However, consideration will need to be given to the geochemistry of the contaminants if they are non-conservative and are being used to provide information on permeability and flow.

Streambed geology and sediment characterisation are required for the effective design of field experiments, hydrological modelling and effective river basin management. In the first instance, this information can come from reconnaissance visits and remote sensing data. However, more in-depth understanding of the processes and sediment sorting patterns is important for hydraulic modelling, because the distribution of sediments influences the boundary roughness and geochemistry. This may require specialist sediment sampling techniques at the site scale or techniques such as airborne LiDAR, in conjunction with geomorphological modelling [32] at the reach or catchment scale. Remote sensing technologies offer the advantage of providing remote access to areas of restricted ground access which potentially offers a new opportunity to start to undertake higher-resolution stream sediment mapping, for example, as undertaken by Miklin and Galia [33]. Geomorphological mapping of this type offers the potential to predict the lateral extent of the hyporheic zone in areas where active channel migration occurs.

3. Analysis

3.1 Hydrological characterisation

Fundamental to researching the hydrology of the hyporheic zone is the development of the conceptual understanding of the host system. A conceptual model is necessary to facilitate targeting of the dominant hyporheic zone processes at the catchment or reach-scale. This requires understanding of the underlying catchment characteristics including land use, the climate and meteorology, geology and geomorphology. Whilst stream hydrology can be considered in terms of catchment recharge, throughflow and discharge (or infiltration, surface run-off, interflow and base-flow), flooding, seasonality and environmental interaction (including anthropogenic factors), extension to the hyporheic zone requires greater consideration of the bed sediments and the hydrological exchange processes therein. For example, in the UK, the Centre for Ecology and Hydrology (CEH) Integrated Hydrological Digital Terrain Model derives outputs for flood modelling based on the range of catchment parameters presented in **Table 2**. Using the CEH approach, it is implicit that the hydrology of soil type (HOST, [34]) parameters embrace the hyporheic zone and that the catchment parameters reflect both the stream hydrology and the

Catchment parameter	Description
AREA	Catchment drainage area
BFI HOST	A base-flow index derived from the hydrological properties of soils as detailed in the HOST dataset, which is based on soil types at a 1-km grid [34]
CENTROID; DPLBAR	Modelling descriptors: centroid of the catchment; distance between model nodes
DPSBAR	Mean drainage path slope as an indicator of catchment steepness
FARL	Potential for flood attenuation by reservoirs and lakes
FPEXT	Extent of the floodplain
LDP	Longest drainage path
PROPWET	A catchment wetness index
SAAR	Average rainfall over a standard period
SMD	Mean soil moisture deficit for the standard period
SPRHOST	Standard percentage run-off of each soil type
URBEXT	Extent of urban and suburban land cover

Table 2.
The primary catchment parameters used for flood estimation by the Centre for Ecology and Hydrology.

DOI: http://dx.doi.org/10.5772/intechopen.85218

stream bedforms at both the catchment and reach scale. However, this model does not allow for geomorphological nuances of the stream sediments and underlying geology that are fundamental to understanding hyporheic exchange. Stream bedforms and their sediments are commonly a product of current, historic and geological climatic conditions. Typically in the UK, drainage patterns and stream sediments reflect the legacy of climatic change throughout the Quaternary, and as a consequence the average bed load grain size may exceed that of the current hydrological conditions of the river or stream, i.e. stream under-fit.

In the UK, sediments deposited in upland catchments are dominated by braided streams that are graded from cobbles in the higher reaches, to gravel and sand in the lower reaches [35, 36]. In lower alluvial floodplains, meandering rivers commonly occupy larger floodplains within which channel migration occurs and the sediments are dominated by sand silt and clay-grade particles. Within each of these environments, stream geomorphology influences the distribution of sediments, e.g. formation of point bars on the inner bank of meander bends. As well as the large-scale lithological variations observed in point bar deposits [37], there are smaller-scale variations, whereby mud-prone sediment is interbedded with sand-prone sediment as inclined heterolithic strata [38]. These complex internal heterolithic variations are particularly important when considering circumstances where chute channels may traverse the point bar deposit because the internal architecture of the hyporheic zone could be more complex and its lateral extent altered. Fine-grained counter point bars may occur on the outer bank of the meander bend where the bend becomes convex in shape [39]. These deposits baffle throughflow due to their fine-grained nature and contrast with eddy-accretion deposits, in turn characterised by thick, sand-prone accumulations [40, 41], which would more easily encourage throughflow. These local variations in permeability affect streambed biogeochemistry and the potential lateral and vertical extents of the hyporheic zone.

The hydraulic connectivity of streams and groundwater can also be considered at the catchment scale and conceptualised in terms of losing and gaining stretches of the stream, where a gaining reach is one that is supplemented by groundwater and a losing reach is one where a proportion of the stream water recharges the underlying aquifer [42]. Perennial streams losing stretches of perennial streams are in constant continuity with groundwater, whereas continuity of gaining stretches may vary with groundwater levels, unless continuity is maintained by switching from gaining to losing conditions. Gaining and losing reaches can be defined from the relationship between stream water level and the potentiometric surface of the groundwater or by changes in the stream discharge, unless subject to artificial modification. The zones of recharge and discharge from the streambed are likely to migrate in accordance with the hydraulic conditions, e.g. due to flooding or seasonality. In many streams the area of discharge or recharge is not evenly distributed across the streambed, which may be due to head differences, differences in hydraulic conductivity or the structure and composition of the streambed sediments.

Hyporheic zone flow paths can be both diffuse and focused. Diffuse flow will reflect the matrix permeability, whereas focused flow may comprise bypass flow that utilises macropores and pipes in the unsaturated zone [43, 44]. The bedrock geology and sediment source zones are reflected in the properties of the hyporheic zone sediments, whilst the form or geomorphology of the sediments reflects the hydraulic setting and its influence on sediment distribution [45]. Flow rates fluctuate around a meander bend, whereby stronger currents are observed at the outer bank and weaker flow is observed on the inner bank [46]. In meandering fluvial systems where alluvium is accreted onto point bar deposits, the finer-grade alluvium is deposited on the inner bank of the downstream limb of the bend, whereas the coarser-grade alluvium is deposited at, and upstream of, the meander apex [37, 47].

The variation in grain size distribution within the sediments gives rise to variability in flow of water through the deposits; more mud-prone sediment has lower porosity and permeability than more sand-prone sediment. The fluctuating flow rates around a meander bend lead to reach-scale variations in sediment-bedded stream hydrological processes commonly leading to the formation of pool and riffle bedforms. Pools are associated with fast, turbulent flow, and consequently the streambed at the pool is comprised of coarse grains up to pebble and cobble calibre [48], which may make these areas proficient in enabling throughflow of water. There is an inverse relationship between both relative pool depth and distance between pools and increasing channel gradient [49]. Generally, the riffles are points of down welling or infiltration at the upstream or stoss side and discharge at the crest or lee side of the riffle [1, 2, 5, 27]. However, the catchment context can result in exceptions to this, e.g. Magliozzi et al. [50].

Streambed permeability is susceptible to modification by other factors, such as the chemical processes that give rise to dissolution or precipitation [51], physical sedimentation and clogging (colmation) and biofilm formation and modification by the streambed fauna. Chen et al. [52] established that influent groundwater flow paths are associated with fine sediment removal from the sediment matrix, thereby increasing the near-surface hydraulic conductivity, whereas downward entrainment of fine particles resulting in siltation or clogging of the sediment in zones of effluent flow (losing reaches of a stream) is associated with a reduction in permeability. For the Nebraskan example described by Chen et al. [52], the alterations to the streambed permeability extended to depths of 5 m or more.

Bypass flow, also termed "preferential" or "fast" flow, is transmitted at orders of magnitude greater than the Darcian matrix flow. Soil pipes are the largest category of macropore and can form connected networks [53]; they have been defined as macropores that are sufficiently large for water to sculpt their form [54]. Soil erosion by throughflow of water gives rise to conduits for lateral or vertical flow [44]. The maximum diameter that a pipe can reach without collapsing will be determined by the density, structure and grain size of the sediment. Piping is most common in the hyporheic zone in bank sediments, locations where surface water is focused by the topography and also where groundwater is focused in gaining reaches of a stream.

The hyporheic zone is related to the base-flow by the residence time, which can be measured using tracers [55]. Investigating a mountain stream at an "Experimental Forest" in Oregon, Haggerty et al. [55] characterised the residence time from a tracer breakthrough curve with a long-tail, poorly characterised by an exponential model and indicative of a large range of exchange timescales, each associated with different volumes of water. This would seem to be representative of many sediment-bedded streams with ranges in permeability and hydraulic regimes. The hyporheic zone also responds to seasonality and importantly for stream ecology, may extend the wet season of ephemeral stream reaches and continue to provide base-flow after the surface stream appears to have "dried up".

Flooding can affect the hyporheic zone causing a hydraulic response as flood waters extend into it. The response is scale dependent and is greater where the channel is unconstrained [56]. In unconstrained systems where channel changes occur, morphological change in the channel may result in localised steepening of the hydraulic gradient or abandonment of hyporheic zones associated with abandoned channels [56, 57]. As the flooding recedes and base-flow re-establishes, the volume of the hyporheic zone will adjust to the new conditions, which has implications for shifts in the stream ecology. During flooding, the increased velocity of surface water can impact the hyporheic zone in accordance with the Venturi Effect. This comprises a net pressure decrease as a function of water velocity, thereby

Factor	Characterising data requirements
Hydrological	Precipitation; temperature; variation with aspect, temporal and seasonal variability (changing spatial and vertical distribution of the hyporheic zone); flooding and catchment parameters (**Table 1**)
Topographical	Slope gradients and lengths, aspect, relationship of sub-catchments to catchments and reach-scale representation
Scale	Regional, catchment, reach or geomorphological feature
Hydrogeological	Head, seepages seasonality as indicated by storativity and head change, permeability and transmissivity at all scales
Geological	Bedrock and superficial lithologies, geological structure, weathering and erosion history, alluvial sediment thickness and architecture and sediment mobility
Valley type	Degree of channel confinement and stream connectivity, upland versus lowland channels, longitudinal and lateral valley gradients, distribution of in-channel bedforms and bank conditions
Vegetation	Types of vegetation: in-channel and riparian and their influence on the hydrology in terms of water balance, hydrogeochemistry and hydrology, e.g. Reynolds number
Urbanisation	Culverted drainage inputs, hydrological changes due to sustainable urban drainage scheme implementation, hard and soft engineering impact on channel characteristics [59] and ecological impacts (biological and chemical)
Climate	Potential climate change impacts, e. g. impacts on existing conditions [60], and climatic zone analogues, e.g. Peel et al. [61]
Biogeochemical and ecohydrological	Chemical and biological characteristics and diversity, nutrient potential, organic matter content and habitats

Table 3.
Factors to consider in catchment scale hyporheic zone modelling.

potentially giving rise to a local reversal in the hydraulic gradient and influx of deeper groundwater [58], despite the higher river stage. Understanding exchanges of this type is particularly important in the characterisation of the hyporheic zone for the management of flood risk.

Hyporheic zone hydrology research scales range from catchment to valley or reach scale. Fully integrated catchment scale modelling and management requires catchment through valley and reach-scale understanding of the hyporheic zone exchange [50]. The key factors underpinning such an analysis correspond with the requirements for conceptual modelling detailed above and summarised in **Table 3**. Each of these factors is subject to scaling issues and time-dependent variability. The range in the scale of hyporheic exchange flows also impacts the attenuating and ecological benefits that are derived from the hyporheic zone in different zones within a catchment. For example, a system dominated by groundwater recharge through coarse sediments may mask the potential hydrochemical benefits afforded by pools and riffles. This complexity is additional to the broader understanding of gaining and losing stretches of the stream or river and is informed by detailed bedform architecture as related to geomorphological processes and described above. Additionally, there are temporal and climatic variations that are fundamental to understanding the functioning of the hyporheic zone.

There are numerous modelling approaches that reflect the multifaceted nature of the hyporheic zone, including ground- and surface water flux and biological and chemical gradients. Hydrological modelling can incorporate the hyporheic zone as a single storage component (e.g. [62]) with fractal scales of response (ranges of permeability and flow path lengths) accounting for the very long tail in the hydrograph

(Haggerty et al., [55]). Alternatively, attempts have been made to model both the relative stationarity of surface water in the main channel and the transient storage (characterised by advection and dispersion) in the stagnant zones separately, e.g. Kazezyilmaz-Alhan and Medina [26].

Further research is required to understand the long-term impacts of environmental change on the hyporheic zone. Undoubtedly the hyporheic zone is susceptible to environmental change (climate and anthropogenic impacts) leading to chemical changes, such as pH change or the addition of contaminants. However, global recognition of the value of the hyporheic zone suggests that, although subject to modification, the value of the hyporheic zone will continue in the context of climate change. For example, it is important in both Arctic streams [27] and tropical streams [63]. Therefore it is likely that the hyporheic zone will be increasingly valued in the context of climate change, because of its buffering capacity (hydrological, chemical and ecological), particularly in the light of forecasts of temperature extremes and higher-intensity rainfall and consequential flooding events.

Increased urbanisation is associated with a decrease in stream base-flow [64]. This is, in part, a consequence of hard engineering, with both consequential reduction in recharge to the hyporheic zone and occlusion of its discharge zones. This in turn, reduces the buffering for flood events, leading to higher flood levels and a need for increased levels of engineering to push flood waters through urban environments as quickly as possible. This urban reduction in stream base-flow reduces the calibre of alluvium accumulating on the streambed, baffling circulation of water between the stream and the hyporheic zone, which in turn reduces the overall water capacity of the system by reducing bank and streambed storage and potentially increasing flood risk. Additionally, increased plastic sedimentation in urban areas may prevent circulation altogether by forming an impermeable barrier. The susceptibility of the ecology of the hyporheic zone to environmental change adds further to the pressure on its hydrological functioning, e.g. algal blooms increase the potential for colmation. The hydrological functioning of the hyporheic zone may also be affected by natural processes such as changes to bedrock weathering and geomorphological processes that affect bank stability. Whilst bedrock susceptibility and geomorphology vary with rock type and hydrological conditions, they tend towards reducing the permeability of the hyporheic zone by the addition of sediment to the streambed, whereas flooding events have the potential to remobilise and transport sediment.

3.2 Hyporheic zones in karst environments

Karst aquifers warrant separate consideration in the context of their hydrological functioning. It is widely recognised that water-mixing zones can be the focus for karst processes within a karst system [65]; therefore, the hyporheic zone is likely to be an important zone within a karstic system. In practice, in many karst aquifers, the hyporheic zone is particularly difficult to define. This is attributed to (i) the difficulty accessing karst systems owing to the range of pore sizes and the difficulty in predicting their distribution, (ii) the rapid change in contact between surface water and groundwater in karst environment, (iii) the water table is commonly poorly defined and (iv) the complexity of some karst systems, e.g. the interplay of matrix and karst porosity in weakly karstic systems, such as chalk [8]. However, the significance of the hyporheic zone in karst systems should not be overlooked particularly with respect to hydraulic function and because of its vulnerability to contamination, attenuation potential and contribution to the evolution of karst systems [65], as well as its ecological value [66].

Karst systems primarily develop their porosity and permeability and therefore their flow paths through dissolution. Dissolution is commonly focused on structural [67], lithological or geochemical boundaries in the lithology, e.g. inception horizons [68]. If, over geological timescales, the hydrological conditions change, the relict karst systems may be abandoned or only partially functioning in the context of the current hydrological regime of a karst aquifer. Furthermore karst systems can form as tiers that are hydrologically focused on different catchment scales, e.g. base level of the major catchment and base level of a sub-catchment, i.e. a losing tributary river. This has the potential to extend the hyporheic zone of a karst system to a considerable depth in some karstic aquifers and makes it difficult to define the groundwater table of karstic aquifers. Characteristically karst aquifers exhibit low storage and therefore seasonally larger changes in head- than matrix-dominated aquifer with higher storage; thus, the hydraulic conditions are variable, and it is difficult to monitor and quantify hyporheic zone processes, particularly when they occur over the very short timescales that are characteristic of more "flashy" catchments.

Humid tropical karst is characterised by limestone hills (towers or pinnacles). Tower karst development appears to be related to the presences of massive crystalline limestone and the development of a system of open, steeply dipping joints that have been exploited by meteoric and shallow groundwater associated with either current or past climatic regimes. Twidale [69] argues that the pinnacles are the consequence of subsurface-weathering fronts and that weathering has been achieved by deep phreatic waters retained in a regolith. Associated with the pinnacles are a succession of notches, comprising sub-horizontal, laterally, solutionally enlarged conduit systems. It is suspected that these are indicative of Pleistocene interglacial high sea levels at elevations of tens of metres above the current sea level, and it has been suggested that they are formed by swamp waters and that subsoil solution may be associated with the formation of cliff foot caves. Without being explicit in the literature, these formational processes are clearly associated with the hyporheic zone. Further evidence of the significance of the hyporheic zone in karst processes comes from progressive increases in the base-flow index towards the lower end of karst streams. Whilst, traditionally, speleogenesis has been conceptualised from the hydrogeological perspective of source, throughflow and discharge with the independent flow paths of the unsaturated zone becoming more ordered at the "water table". Diffusely recharged water in karst aquifers with long residence times is quickly saturated by carbonate [65], even with the addition of carbon dioxide derived from vegetation or the pH change attributable to overlying acid-generating soils, suggesting that an additional process is required for dissolutional enlargement at the discharge end of the system. Gulley et al. [70] in their investigation of the less mature karst systems of the Suwannee River in North-Central Florida established that undersaturated floodwater-related dissolution during flow reversal in the hyporheic zone of karst discharge areas likely accounts for a significant component of the dissolutional enlargement at the downstream end of the flow path. Additionally, exchange between different scales of conduit contributes to attenuation of contaminants in this zone. More specifically, flood water-dispersed contaminants will be displaced by distal groundwater as the floodwater recedes, thereby contributing to dilution.

Flood waters are also responsible for the delivery of sediment via the hyporheic zone into karst systems, and this has a significant role in the armouring of portions of the karst aquifer and attenuation of contaminants [71], but is prone to disturbance by subsequent flooding. Similarly, algal armouring of conduit surfaces is sustained by nutrient supplied by hyporheic exchange. The processes that take place in the hyporheic zone of karst aquifers occur at a range of scales from that of a conduit-wall topography, e.g. scallops to large-scale conduit networks. Arguably, rapid exchange over shorter flow paths [72] is probably the most hydrologically significant type of hyporheic zone impact on karst aquifers, because of the greater potential for dissolution resulting from rapid water exchange. However, this is a relatively new research area, and interplay with delivery of soils gases and carbon storage (e.g. [73, 74]) may place this in a different perspective in the longer term.

Two additional aspects of the hyporheic zone have evolved from karst research. Firstly, the relative independence of karst flow paths has led some authors to consider karst conduits as being analogous to surface streams and therefore suggest that karst conduit hyporheic zones can be characterised (e.g. [75]). Whilst this is undoubtedly true because of the influence on biology, contaminant attenuation, geochemistry and speleogenesis of karst systems, such an approach is likely to lead to unnecessary ambiguity in hyporheic zone research. Secondly, the extensive research on the use of tracers to identify flow paths in karst aquifer provides a rich resource in terms of understanding the availability, benefits and challenges associated with a broad range of techniques, e.g. Smart [28].

4. Results from case studies

The two case studies presented below comprise accounts of research that have been undertaken by the British Geological Survey to further understanding of the hydrogeochemical functioning and value of the hyporheic zone in the context of groundwater contamination. Other British Geological Survey research projects include research on the role of the hyporheic zone in flooding in Oxford [76] and Lambourne [8] in Southern England and Eddleston, Scotland [77].

4.1 Rookhope Burn, wear sub-catchment of Northumbria River Basin District, Northern England

This case study was focused on the Rookhope Burn, an upland stream forming a tributary of the River Wear in northern England. Here the potential for attenuation in the hyporheic zone was explored because the stream lies within a significantly mine-impacted area of the North Pennines Orefield, UK. Zinc had been identified as the contaminant of concern. Mass balances of in-stream and in-flow (subsurface, historic mine working related, contributions of zinc-contaminated groundwater) chemical loads determined from major and trace elements concentrations and synoptic flow monitoring had identified sinks as well as sources of zinc in the burn. The sources comprise rising, mining-contaminated groundwater [78] with the potential to shift the hyporheic zone [56].

In order to investigate this further, the physicochemical composition of the hyporheic zone, a stream stretch, was characterised at two contrasting flow and temperature regimes [13]. The Rookhope Burn streambed comprises coarse-textured river terrace deposits. The underlying bedrock geology is formed of mineralised Dinantian limestones capped by Namurian sandstones and mudstones. For this catchment vertical element concentration gradients were obtained using multilevel samplers down to a depth of 40 cm below the water-sediment boundary and along a 12-m reach. Additionally, in situ diffuse gradients in thin film (DGT) measurements of surface water and porewater were obtained (**Figure 1**).

The multilevel samplers described by Dearden and Palumbo-Roe [18], like those used at Polmadie Burn and described below, comprised a 12 mm ID 1200 mm long, HDPE pipe, fitted at one end with a stainless steel drive-point to enable penetration of the device into sediments. The pipe had two 4-mm-diameter holes at the base

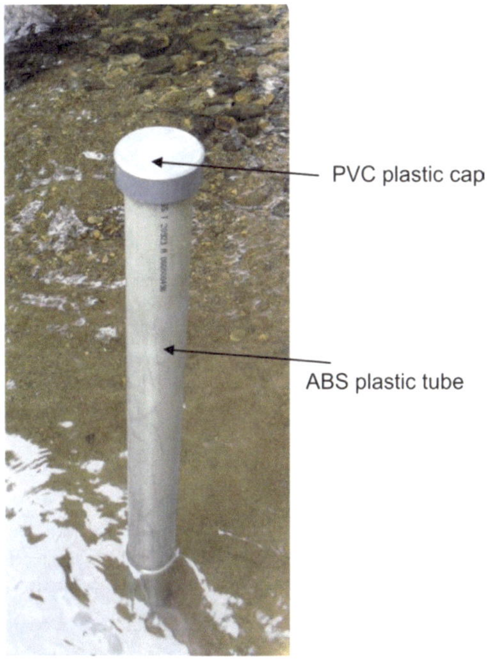

Figure 1.
DGT placed in the base of the hollow sampler and left for 24 hours.

to allow the piezometric surface within the hyporheic zone to be monitored. Four discrete 1.6 mm ID Teflon sampling tubes were attached around the central pipe and were terminated such that ports were located at 10-cm intervals. Each of the ports was encased in (45 μm) nylon mesh screen to prevent sediment blockages. Porewater samples were recovered using a low-flow multichannel peristaltic pump that enabled simultaneous sampling of the four ports at an approximate flow rate of 4 mL/min; pH and Eh were measured using a flow-through cell, and Pt electrode Eh-measured values were corrected to the standard hydrogen electrode. The level of the hyporheic zone water was measured relative to the river water level using a mini dipper placed down the central HDPE tube [13].

Two sampling surveys were undertaken, one in the summer (July 2010) and one in the autumn (October 2010). The hydrological conditions in the catchment were very consistent for at least 10 days before each sampling event. In addition to the multilevel samples, samples of surface water were collected at each of the four locations. Groundwater was also collected from a private well adjacent to the sampling site for comparison with the hyporheic porewaters [13].

Near neutral pH and oxidising conditions characterised the hyporheic zone. The upper 15–20 cm was dominated by the chemistry of the overlying water, whilst interactions with the solid phase occurred in the deeper part of the hyporheic zone. Mineralogical analysis of manganese-rich grain coatings from the bed sediment indicated that manganese was being attenuated in the hyporheic zone. Additionally, there was clear evidence for hyporheic porewater enrichment in lead that was unaffected seasonally, whilst zinc concentrations were higher in July. The significance of the observed sediment-scale hyporheic processes on the reach-scale geochemical mass balance was estimated by using surface water geochemical loading calculations. Along a 700-m stream stretch of the burn, a constant loss of manganese stream load and continuous gain of lead stream load with more temporally variable zinc stream loading were measured. This demonstrated that the hyporheic zone of the mine-impacted stream supports steep manganese, lead and zinc gradients. Seasonality in the hyporheic zone was suspected from the results, but was not fully investigated, and therefore further investigation of seasonality was identified as a necessary future research direction for this catchment. More specifically, this would require periods of continuous monitoring, which is difficult to establish in remote catchments.

4.2 Polmadie Burn, Clyde catchment, Scotland

The potential for hyporheic zone attenuation of chromium contamination in Polmadie Burn, a minor tributary of the River Clyde in Scotland, UK, was investigated during two monitoring periods in February and September 2012 [79]. A summary of the monitoring techniques and findings are presented here to demonstrate the importance of linking hydrology to geochemical assessments of the context of hyporheic zone.

The Polmadie Burn is an urban steam, in the order of 3 km to the south-east of Glasgow City Centre. It is impacted by hexavalent chromium-rich effluent leached from the landfilled residue from historical (late nineteenth and early twentieth century) processing of chromite ore [80]. The stream responds rapidly to changes in the discharge from its culverted urban drainage catchment. The underlying Carboniferous bedrock comprises cyclic sequences of mudstone, siltstone and sandstone that are overlain by superficial deposits of river terrace deposits capped by low-permeability alluvium. The low permeability of the hyporheic zone contrasts with the coarse streambed investigated at Rookhope.

Following a desk study investigation of the burn, a sampling strategy was designed to integrate hydrology and geochemistry in the context of spatial and temporal change. A specific aspect of the investigation was the attention given to the health and safety issues associated with working in the chromium-contaminated stream with a soft silty streambed (**Figure 2**). Upstream and downstream transects were selected for sampling, which included in situ monitoring via multilevel mini-point piezometers (**Figure 2**), hydraulic testing (falling head slug tests in the bank sediments) as well as sampling and geochemical/mineralogical characterisation of hyporheic water, surface water and streambed sediments. At each transect the multilevel piezometers comprised a number of carefully labelled sampling ports to facilitate hydraulic gradient determination and vertical profile sampling from within the bed sediments at depths of up to 90 cm below bed level. Following a stabilisation period of 12 hours, sampling was undertaken with a low-flow multichannel peristaltic pump. The fieldwork comprised two campaigns to monitor the hyporheic zone processes in operation at different stages of the stream. The February visit allowed sampling of water from the hyporheic zone and stream over a large short-term variation in stream depth, which comprised a low tide during which there was temporary exposure of river bed sediments followed by overnight rainfall and high stream levels. The second field visit, in September, coincided with more constant hydraulic conditions. This allowed synoptic surface water quality sampling to be carried out to provide qualitative evidence of any whole-stream-contaminant attenuation to which the hyporheic zone may contribute.

Detrital grains of historical chromium process residue were found to contribute to the total chromium concentrations (size fraction < 150 μm) that reached 8800 mg kg^{-1} in the streambed sediment. There was a sharp decrease of total dissolved (filtered < 0.45 μm) chromium concentrations at the surface water-sediment boundary in all profiles, from a mean chromium concentration of 1100 μg l^{-1} in the surface water to 5 μg l^{-1} in the porewater. This was associated with an elevated ferrous iron concentration in the porewater (mean concentration 1700 μg l^{-1}) and chromium (VI) reduction to chromium (III) solids of low solubility. However, the hyporheic zone did not respond to the large short-term changes of stream stage as indicated by

Figure 2.
Hyporheic zone multilevel sampling of the bed of Polmadie Burn, Glasgow. Note how, for ease of access, the multipoint sampling tubes were extended from the sampling point to the base station where the peristaltic pump was deployed.

the hyporheic zone water composition. It was concluded that the low-permeability alluvial sediments imposed a limit on the effectiveness of the hyporheic zone for enhancing Cr surface water quality at the reach scale. This was also evident in the surface water quality synoptic sampling which showed only moderate to low downstream decreases in surface water chromium concentrations. Thus, the key finding was that although the geochemical potential for hyporheic attenuation of surface water chromium was clearly established, the hydraulic functioning of the hyporheic zone was limited by poor hydrological connectivity with the Polmadie Burn.

Acknowledgements

This chapter is published with the permission of the Executive Director of the British Geological Survey (UKRI). Stephanie Bricker (British Geological Survey) and an unnamed reviewer are thanked for insightful reviews and suggestions leading to improvements of the manuscript.

Author details

Vanessa J. Banks[1*], Barbara Palumbo-Roe[1] and Catherine E. Russell[2]

1 British Geological Survey, Environmental Science Centre, Keyworth, UK

2 University of Leicester, UK

*Address all correspondence to: vbanks@bgs.ac.uk

IntechOpen

© 2019 The Author(s). Licensee IntechOpen. This chapter is distributed under the terms of the Creative Commons Attribution License (http://creativecommons.org/licenses/by/3.0), which permits unrestricted use, distribution, and reproduction in any medium, provided the original work is properly cited.

References

[1] Cardenas MB. Hyporheic zone hydrologic science: A historical account of its emergence and a prospectus. Water Resources Research. 2015;**51**:3601-3616

[2] Bencala KE, Walters RA. Simulation of solute transport in a mountain pool-and-riffle stream: A transient storage model. Water Resources Research. 1983;**19**:718-724

[3] Mueller ER, Pitlick J. Sediment supply and channel morphology in mountain river systems: 1. Relative importance of lithology, topography, and climate. Journal of Geophysical Research: Earth Surface. 2013;**118**:2325-2342

[4] Orghidan T. Ein neuer lebensraum des unterirdischen wassres der hyporheischen biotope. Archiv für Hydrobiologie. 1959;**55**:392-414

[5] Smith JWN. Groundwater-surface water interactions in the hyporheic zone. In: Science Report SC030155/SR1. Environment Agency; 2005. 65 pp

[6] Brunke MA, Gonser T. The ecological significance of exchange processes between rivers and groundwater. Freshwater Biology. 1997;**37**:1-33

[7] Krause S, Fleckenstein J, Hannah D, Heppell K, Kaeser D, Pickup R, et al. Inter-disciplinary perspectives on processes in the hyporheic zone. Ecohydrology. 2011;**4**:481-499

[8] Lapworth DJ, Gooddy DC, Allen D, Old GH. Understanding groundwater, surface water, and hyporheic zone biogeochemical processes in a Chalk catchment using fluorescence properties of dissolved and colloidal organic matter. Journal of Geophysical Research. 2009;**114**:G00F02. 10 pp

[9] Bricker SH, Banks VJ, Galik G, Tapete D, Jones R. Accounting for groundwater in future city visions. Land Use Policy. 2017;**69**:618-630

[10] Boulton AJ, Datry T, Kasahara T, Mutz M, Stanford JA. Ecology and management of the hyporheic zone: stream-groundwater interactions of running waters and their floodplains. Journal of the North American Benthological Society. 2010;**29**(1):26-40

[11] Peralta-Maraver I, Reiss J, Robertson AL. Interplay of hydrology, community ecology and pollutant attenuation in the hyporheic zone. Science of the Total Environment. 2018;**610-611**:267-275

[12] White DS. Perspectives on defining and delineating hyporheic zones. Journal of the North American Benthological Society. 1993;**12**(1):61-69

[13] Palumbo-Roe B, Dearden R. The hyporheic zone composition of a mining-impacted stream: Evidence by multilevel sampling and DGT measurements. Applied Geochemistry. 2013;**33**:330-345

[14] Buss S, Cai Z, Cardenas B, Fleckenstein J, Hannah D, Heppell K, et al. The Hyporheic Handbook: A Handbook on the Groundwater-Surface Water Interface and Hyporheic Zone for Environment Managers. Bristol, UK: Environment Agency; 2009; 280 pp. (Science Report, SC0500)

[15] Rosenberry DO. A seepage meter designed for use in flowing water. Journal of Hydrology. 2008;**359**:118-130

[16] Burk L, Cook PG. A simple and affordable system for installing shallow drive point piezometers. Technical Note. Groundwater Monitoring and Remediation. 2015;**35**(3):101-104

[17] Byrne P, Zhang H, Heathwaite L, Binley A, Ullah S, Kaeser D, et al.

Passive pore water sampling provides evidence of suppression of hyporheic exchange and nitrate transformation in a groundwater-fed river. In: Proceedings of the 13th International Conference on Environmental Science and Technology; Athens, Greece; 5-7 September 2013

[18] Dearden R, Palumbo-Roe B. Technical note: Hyporheic zone sampling procedures. In: Groundwater Science Programme Open Report OR/10/048. British Geological Survey; 2010. 21 pp

[19] Bradley DC, Omerod SJ. Evaluating the precision of kick-sampling in upland streams for assessments of long-term change: The effects of sampling effort, habitat and rarity. Archiv fűr Hydrobiologie. 2002;**155**(2):199-221

[20] Stubbington R, Dole-Olivier M-J, Galassi DMP, Hogan J-P, Wood PJ. Characterization of Macroinvertebrate Communities in the Hyporheic Zone of River Ecosystems Reflects the Pump-Sampling Technique Used. PLoS ONE. 2016;**11**(10):e0164372. https://doi.org/10.1371/journal.pone.0164372

[21] Ward AS, Gooseff MN, Singha K. Imaging hyporheic zone solute transport using electrical resistivity. Hydrological Processes. 2010;**24**:948-953

[22] Shaw EM. Hydrology in Practice. 3rd ed. London: Chapman & Hall; 1994. 569 pp

[23] Dogwiler T, Wicks C. Thermal variations in the hyporheic zone of a karst stream. International Journal of Speleology. 2006;**35**(2):59-66

[24] Gerecht KE, Cardenas MB, Guswa AJ, Sawyer AH, Nowinski JD, Swanson TE. Dynamics of hyporheic flow and heat transport across a bed-to-bank continuum in a large regulated river. Water Resources Research. 2011;**47**:W03524. 12 pp

[25] Marzadri A, Tonina D, McKean JA, Tiedemann MG, Benjankar RM. Multi-scale streambed topographic and discharge effects on hyporheic exchange at the stream network scale in confined streams. Journal of Hydrology. 2014;**519**:1997-2011

[26] Kazezyilmaz-Alhan CM, Medina MA. Stream solute transport incorporating hyporheic zone processes. Journal of Hydrology. 2006;**329**:26-38

[27] Edwardson KJ, Bowden WB, Dahm C, Morrice J. The hydraulic characteristics and geochemistry of hyporheic and parafluvial zones in Arctic tundra streams, north slope, Alaska. Advances in Water Resources. 2003;**26**:907-923

[28] Smart CC. Artificial tracer techniques for the determination of the structure of conduit aquifers. Groundwater. 1988;**26**(4):445-453

[29] Cranswick RH, Cook PG, Lamontagne S. Hyporheic zone exchange fluxes and residence times inferred from riverbed temperature and radon data. Journal of Hydrology. 2014;**519**:1870-1881

[30] Möller P, Dulski P, Bau M, Knappe A, Pekdeger A, Sommer-von Jarmersted C. Anthropogenic gadolinium as a conservative tracer in hydrology. Journal of Geochemical Exploration. 2000;**69-70**:409-414

[31] Ciszewski D, Aleksander-Kwaterczak U. Contrasting sediment and water chemistry indicates the extent of the hyporheic zone in a polluted river system. Geology, Geophysics and Environment. 2016;**42**:151-159

[32] Jones AF, Brewer PA, Johnstone E, Macklin MG. High-resolution interpretative geomorphological mapping of river valley environments using airborne LiDAR data. Earth Surface Processes and Landforms. 2007;**32**:1574-1592

[33] Miklin J, Galia T. Detailed fluvial-geomorphological mapping of wadeable streams: A proposal of universal map symbology. Journal of Maps. 2017;**13**:698-706

[34] Boorman DB, Hollis JM, Lilly A. Hydrology of Soil Types: A Hydrologically-Based Classification of the Soils of the United Kingdom. IH Report No. 126. Wallingford: Institute of Hydrology; 1995

[35] Paola C, Parker G, Seal R, Sinha SK, Southard JB, Wilcock PR. Downstream fining by selective deposition in a laboratory flume. Science. 1992;**258**:1757-1760

[36] Powell DM. Patterns and processes of sediment sorting in gravel-bed rivers. Progress in Physical Geography. 1998;**22**(1):1-32

[37] Russell CE, Mountney NP, Hodgson D, Colombera L. A novel approach for prediction of lithological heterogeneity in fluvial point-bar deposits from analysis of meander morphology and scroll-bar pattern. International Association of Sedimentologists Special Publication. 2018;**48**

[38] Thomas RG, Smith DG, Wood JM, Visser J, Calverley-Range EA, Koster EH. Inclined heterolithic stratification—Terminology, description, interpretation and significance. Sedimentary Geology. 1987;**53**:123-179

[39] Smith DG, Hubbard S, Leckie D, Fustic M. Counter point bars in modern meandering rivers: Recognition of morphology, lithofacies and reservoir significance, examples from Peace River, AB, Canada. Sedimentology. 2009;**56**:1655-1669

[40] Burge LM, Smith DG. Confined meandering river eddy accretions: Sedimentology, channel geometry and depositional processes. In: Smith ND, Rogers J, editors. Fluvial Sedimentology VI. Vol. 28. International Association of Sedimentology, Special Publication; 1999. pp. 113-130

[41] Smith DG, Hubbard SM, Lavigne J, Leckie DA, Fustic M. Stratigraphy of counter-point-bar and eddy-accretion deposits in low-energy meander belts of the peace-Athabasca Delta, Northeast Alberta, Canada. In: Davidson SK, Leleu S, North CP, editors. From River to Rock Record: The Preservation of Fluvial Sediments and Their Subsequent Interpretation. SEPM; 2011. p. 143, 152. Special Publication 97

[42] Sophocleous M. Interactions between groundwater and surface water: The state of the science. Hydrogeology Journal. 2002;**10**:52-67

[43] Kirby M. Hillslope runoff processes and models. Journal of Hydrology. 1988;**100**:315-339

[44] Wilson GV, Wells RR, Kuhnle RA, Fox GA, Nieber J. Sediment detachment and transport processes associated with internal erosion of soil pipes. Earth Surface Process and Landforms. 2018;**43**(1):45-63

[45] Church M. Bed material transport and the morphology of alluvial river channels. Annual Review of Earth and Planetary Sciences. 2006;**34**:325-354

[46] Jackson RG. Depositional model of point bars in the lower Wabash River. Journal of Sedimentary Petrology. 1976;**46**:579-594

[47] Fustic M, Hubbard SM, Spencer R, Smith DG, Leckie DA, Bennett B, et al. Recognition of down-valley translation in tidally influenced meandering fluvial deposits, Athabasca Oil Sands (Cretaceous), Alberta, Canada. Marine and Petroleum Geology. 2012;**29**:219-232

[48] Allen JRL. A review of the origin and characteristics of recent

alluvial sediments. Sedimentology. 1965;**5**:89-191

[49] Wohl EE, Vincent KR, Merritts DJ. Pool and riffle characteristics in relation to channel gradient. Geomorphology. 1993;**6**:99-110

[50] Magliozzi C, Grabowski R, Packman AI, Krause S. Scaling down hyporheic exchange flows: From catchments to reaches. Hydrology and Earth System Sciences. 2016. DOI: 10.5194/hess-2016-683

[51] Fuller CC, Harvey JW. Reactive uptake of trace metals in the hyporheic zone of a mining contaminated stream, Pinal Creek, Arizona. Environmental Science and Technology. 2000;**34**:1150-1155

[52] Chen X, Dong W, Ou G, Wang Z, Liu C. Gaining and losing stream reaches have opposite hydraulic conductivity distribution patterns. Hydrology and Earth System Sciences. 2013;**17**:2569-2579

[53] Jones JAA. Soil piping and catchment response. Hydrological Processes. 2010;**24**:1548-1566

[54] Beven K, Germann P. Macropores and water flow in soils. Water Resources Research. 1982;**18**(5):1311-1324

[55] Haggerty R, Wondzell SM, Johnson MA. Power-law residence time distribution in the hyporheic zone of a 2nd-order mountain stream. Geophysical Research Letters. 2002;**29**(13):1640-1644

[56] Wondzell SM, Swanson FJ. Floods, channel change, and the hyporheic zone. Water Resources Research. 1999;**35**(2):555-567

[57] Boano F, Revelli R, Ridolfi L. Reduction of the hyporheic zone volume due to the stream-aquifer interaction. Geophysical Research Letters. 2008;**35**:L09401. 5 pp

[58] Sickbert T, Peterson EW. The effects of surface water velocity on hyporheic interchange. Journal of Water Resource and Protection. 2014;**6**:327-336

[59] Zhou T, Endreny TA. Reshaping of the hyporheic zone beneath river restoration structures: Flume and hydrodynamic experiments. Water Resources Research. 2013;**49**:5009-5020

[60] Zhou S, Xingzhong Y, Peng S, Yue J, Xianofeng W, Liu H, et al. Groundwater-surface water interactions in the hyporheic zone under climate change scenarios. Environmental Science and Pollution Research. 2014;**21**(24):13943-13955

[61] Peel MC, Finlayson BL, McMahon TA. Updated world map of the Köppen-Geiger climate classification. Hydrology and Earth System Sciences. 2007;**11**:1633-1644

[62] Choi J, Harvey JW, Conklin M. Characterizing multiple timescales of stream and storage zone interaction that affect solute fate and transport in streams. Water Resources Research. 2000;**36**(6):1511-1518

[63] Covich AP, Crowl TA, Scatena FN. Effects of extreme low flows on freshwater shrimps in a perennial tropical stream. Freshwater Biology. 2003;**48**:1199-1206

[64] Paul MJ, Meyer JL. Streams in the urban landscape. Annual Review of Ecology and Systematics. 2001;**32**:333-365

[65] Dreybrodt W. Mixing corrosion in $CaCO_3$–CO_2–H_2O systems and its role in the karstification of limestone areas. Chemical Geology. 1981;**32**:221-236

[66] Maurice L, Bloomfield J. Stygobitic invertebrates in groundwater—A review from a hydrogeological perspective. Freshwater Reviews. 2012;**5**:51-71

[67] Rugel K, Golladay SW, Rhett Jackson C, Rasmussen TC. Delineating groundwater/surface water interaction in a karst watershed: Lower Flint River Basin, southwestern Georgia, USA. Journal of Hydrology: Regional Studies. 2016;**5**:1-19

[68] Lowe DJ, Gunn J. Carbonate speleogenesis: An inception horizon hypothesis. Acta Carsologica. 1997;**XXVI/2**(38):457-488

[69] Twidale CR. Origin and development of *Karstinselberge*, with particular reference to some South East Asian evidence. Geological Society of Malaysia Bulletin. 2006;**49**:145-155

[70] Gulley J, Martin J, Screaton EJ, Moore P. River reversals into karst springs: A model for cave enlargement in eogenetic karst aquifers. Geological Society of America Bulletin. 2011;**123**(3/4):457-467

[71] Guo F, Wang W, Jiang G, Huang S. Distribution and stable isotopic compositions of organic carbon in surface sediments in hyporheic zone of karst springs. Environmental Earth Sciences. 2016;**75**:850. 13 pp

[72] Henry K, Wilson JL. Relative importance of karst-conduit hyporheic zones co-occurring at different spatial scales. In: American Geophysical Union Fall Meeting 2014. 2014. Abstract only

[73] Gulley J, Martin J, Moore P. Vadose CO_2 gas drives dissolution at water tables in eogenetic karst aquifers more than mixing dissolution. Earth Surface Processes and Landforms. 2014;**39**:1833-1846

[74] Zhou G, Huang J, Tao X, Luo Q, Zhang R, Liu Z. Overview of 30 years of research on solubility trapping in Chinese karst. Earth-Science Reviews. 2015;**146**:183-194

[75] Koski K, Wilson JL. Is the hyporheic zone a useful concept for karst aquifers with conduits? Geological Society of America Abstracts with Programs. 2009;**41**(7):400

[76] MacDonald D, Dixon A, Newell A, Hallaways A. Groundwater flooding within an urbanised flood plain. Journal of Flood Risk Management. 2012;**5**(1):68-80

[77] Ó Dochartaigh BÉ, Archer NAL, Peskett L, MacDonald AM, Black AR, Auton CA, et al. Geological structure as a control on floodplain groundwater dynamics. Hydrogeology Journal. 2018:1-14. DOI: 10.1007/s10040-018-1885-0

[78] Banks VJ, Palumbo-Roe B. Synoptic monitoring as an approach to discriminating between point and diffuse source contributions to zinc loads in mining impacted catchments. Journal of Environmental Monitoring. 2010;**12**:1684-1698

[79] Palumbo-Roe B, Banks VJ, Bonsor HC, Hamilton EM, Watts MJ. Limitations on the role of the hyporheic zone in chromium natural attenuation in a contaminated urban stream. Applied Geochemistry. 2017;**83**:108-120

[80] Farmer JG, Graham MC, Thomas RP, Licona-Manzur C, Paterson E, Bewley RJF. Assessment and modelling of the environmental chemistry and potential for remediative treatment of chromium-contaminated land. Environmental Geochemistry and Health. 1999;**21**(4):331-337

Chapter 3

Accuracy of Hydrogeological Calculations and Forecasts

Mikhail M. Burakov

Abstract

Aquifer systems most often appear to be double-layered or multi-layered. The parameters of groundwater flow from adjacent horizons to the tested ones through separating low-permeable layers or the leakage of groundwater from the low-permeable overlapping layers are determined by the results of pumping. There are methods for determining the permeability parameters of tested horizons and flow parameters by the results of such pumping. However, the issue of assessment of flow parameter confidence remains current. This chapter proposes a method for performing such assessment. The method was tested on a specific example. The obtained error estimates for the parameters of a layered aquifer system are typical for groundwater filtration schemes in aquifers with overflow.

Keywords: aquifers, multilayer aquifer systems, geofiltration parameters, flow parameters through separating low-permeable sediments, accuracy of parameter determination

1. Introduction

There are quite a lot of examples of recording of groundwater (GW) flow from adjacent aquifers of layered formations in estimating the commercial reserves of deposits in artesian basins of various orders, solving other problems of GW filtration. The confidence of such assessments is entirely determined by two main factors, namely by the compliance of the design scheme with the natural conditions and the confidence of the established parameters of the main and adjacent aquifers and the flow parameters. All this shows the urgent need, on the one hand, of assessing the flow values only by direct methods, and on the other, revising the methods of experimental determination of the flow parameters in layered formations with assessments of random errors of such parameters.

2. Assessment of confidence of flow parameters determined by the results of pumping from the layered aquifer systems

2.1 Physical and mathematical models of groundwater filtration in multilayer systems with overflow

As is known, when conducting experimental filtration testing (EFT) of the layered formation to determine the parameters of low-permeable separating

IntechOpen

sediments, the data on the decrease in the GW level mainly in the aquifer tested are used. At the same time, with a decrease in pressure in this aquifer during the process of cluster pumping test, two options of adjacent aquifer reaction are possible: (a) a decrease in the level determined by the pumping does not occur and (b) a decrease in pressure is recorded. In some cases, the elastic filtration in separating layers is significantly manifested.

GW filtration scheme, for which there is no reaction in the adjacent horizon during the EFT, GW movement in the tested horizon to the disturbing well operating with a constant flow rate, with an elastic filtration mode in separating low-permeable sediments, is described based on equation [1]. For a period of time that meets the condition

$$t < \frac{m_0 \mu_0{}^*}{10 k_z},$$

this equation is presented as

$$S = \frac{Q}{4\pi T} H(u,\beta), \quad u = \frac{r^2}{4\chi t}, \quad \beta = \frac{r}{4B}\sqrt{\frac{\mu_0{}^*}{\mu^*}}, \quad B = \sqrt{\frac{m_0 T}{k_z}}, \tag{1}$$

where S is a decrease in piezometric level of GW at a distance r from the disturbance centre at a moment of time t from the start of pumping; Q is the constant flow rate of disturbance; $H(u,\beta)$ is the improper integral tabulated in [1]; $T = Km$, μ^*, K and m are water transmissibility, elastic water yield, permeability factor and the thickness of tested horizon, respectively; $\chi = \frac{T}{\mu^*}$ is the piezoconductivity of the tested horizon; k_z and $\mu_0{}^*$ are permeability factor and elastic water yield of separating low-permeable sediments, respectively; B is the flow factor.

In case of prolonged disturbances, when the estimated time values meet the following condition:

$$t > \frac{5 m_0 \mu_0{}^*}{k_z},$$

Eq. (1) is as follows [1]:

$$S = \frac{Q}{4\pi T} W\left(u\delta_1, \frac{r}{B}\right), \quad \delta_1 = 1 + \frac{\mu_0{}^*}{3\mu^*}, \tag{2}$$

where $W\left(u\delta_1, \frac{r}{B}\right)$ is the well function for layered systems with the flow. The rest designations remain the same.

A special case arising from (2) is the equation obtained in [2], in which the volume of water in low-permeable sediments is assumed to be negligible, that is, $\mu_0{}^* \to 0$, and the filtration mode in low-permeable layer becomes hard. It is presented as

$$S = \frac{Q}{4\pi T} W\left(u, \frac{r}{B}\right). \tag{3}$$

Here, as in (2), $W\left(u, \frac{r}{B}\right)$ is the improper integral tabulated in [3] (the Hantush function). The rest designations remain the same.

Eq. (3) is applicable only to aquifers with relatively small thickness [1], so that the following condition should be fulfilled:

$$\frac{m}{B} \leq 0.1.$$

This assumption is usually fulfilled [4], so accordingly, Eq. (3) is applicable for interpreting the EFT results in most practical cases.

Thus, to date, there is a well-developed theoretical base that provides the fundamental possibility of describing and studying GW filtration in layered systems with GW flow to the tested aquifers from adjacent ones. Based on this, the methods for the interpretation of the results of experimental cluster pumping from the layered aquifer systems were developed in order to determine the parameters of the tested horizon and the parameters of separating low-permeable sediments.

2.2 Formulation of problem of determining the random errors of the parameters

When conducting EFT, the parameters of the tested horizon and separating low-permeable sediments are not measured directly; they are calculated by equations, and these equations include directly measured characteristics of initiated disturbance of the aquifer system as arguments. The characteristics are measured with some errors so that in the parameters of the conducting medium determined by the results of experimental works, all possible errors of measurements of initial data and intermediate calculations appear.

In full accordance with the theory of errors in measurements and calculations in the process and interpretation of the results of test pumping from wells, there is a probability of distortion of the parameters to be determined (water transmissibility T, piezoconductivity χ and elastic water yield μ^* of the tested horizon and flow factor B and permeability factor of the separating low-permeable layer k_z), that is, introducing systematic and random errors [5–11].

Identification and elimination of systematic errors are an integral part of engineering calculations. The main approach to eliminate such errors is to make signed corrections in the results of measurements and calculations. The overwhelming majority of publications on the GW dynamics are devoted to estimates of systematic errors and methods of their accounting; in those publications, all types of errors are studied in sufficient detail.

The situation is different with estimates of random errors of the parameters of aquifer systems to be determined. The latter is due to the cumulative manifestation of many factors, and the nature of the manifestation of each of these factors is not exactly reproduced in the repeated (and subsequent) testing of aquifer systems. Among random errors, the most significant is individual, instrumental, methodological, and model errors [8, 12]. These errors, unlike systematic ones, cannot be excluded from the results of measurements and calculations; therefore, their identification and estimation of the values of such errors are extremely important because, ultimately, they determine the confidence and reliability of the parameters to be set.

However, the issues of assessing the accuracy of primary measurements and the parameters of aquifer systems have not been studied enough. There are several works in the Russian language [9, 13–15], which basically represent the full list of publications with the reported results of the assessment of random, mainly instrumental, errors of the most important parameters, namely water transmissibility, piezoconductivity, and water yield of tested aquifers. In the publications mentioned, the accuracy of the primary measurements and the probable random errors of the parameters were analysed with reference to the methodological recommendations for interpreting the results of pumping from [16–18]. The main disadvantage of the latter is the use of single measurements of the level decrease in wells, which almost always leads to significant systematic methodological errors of the established characteristics, which cannot be identified and eliminated. Estimates of random error parameters in such cases lose their significance.

Therefore, as a standard method for interpreting the EFT results of aquifers and layered systems, the graphic-analytical method, as well as reference curve method, which are largely devoid of this drawback [11, 19], were recommended. Accordingly, in some works [5, 12, 20–22], methodological approaches and results of assessing the reliability of parameters of aquifers, as well as layered aquifer systems with the flow, were described. In the works [23, 24], the results of assessing the individual random errors of the determined parameters and the detection of gross measurement errors are presented.

A well-known common equation is used to assess the confidence of EFT results in aquifers. Thus, if a differentiable function $y = f(x_1, x_2, ..., x_n)$is given, and if mean square error of x_i arguments σ_{x_i} (i = 1, 2, ..., n) is known (where n is the number of arguments), then the mean square error of this function is [7–10, 25]

$$\sigma_y \cong \sqrt{\sum_{i=1}^{n} \left(\frac{\partial y}{\partial x_i} \sigma_{x_i} \right)^2}. \tag{4}$$

Measurement errors in most geological and geographical studies are estimated as limiting [6, 9], as well as in documents on measuring instruments used in hydrogeological practice, maximum permissible random errors are indicated. The latter is determined by the worst conditions that have arisen during measurements of any characteristic of the disturbance: all components of the errors are maximum in absolute value, and all of them are of the same sign. The probability of these errors is fixed and very small, and meanwhile, the need for error estimates with significantly higher probabilities often arises, since the actual errors of measurements and calculations are noticeably less than the limiting ones. Therefore, the guidelines on metrology (as, for example [7–10, 25]) recommend performing the calculations based on Eq. (4), which has a clearly expressed probabilistic nature. At the same time, for a normally distributed quantity, all spreads with an accuracy of fractions of a percent are within 3σ.

This implies that on the basis of (4), one can find out the maximum permissible errors of the parameters; however, for practical calculations, they are usually limited to the confidence probability of 0.954, and then the errors σ_{x_i} with a confidence probability of 0.683 are doubled [7, 9, 10, 25, 26]. Limiting permissible errors of measuring devices and primary measurements correspond to a confidence probability level of 0.997 and constitute $3\sigma_{x_i}$ [7, 9, 10, 25, 26]. Random errors of permeability parameters are estimated lower with a probability of 0.954.

2.3 Assessment of random error parameters of aquifer tested by test pumping

Let us consider the assessment of random errors of parameters of tested aquifer being a part of the multi-layered system, parameters of separating low-permeable sediments for the filtration scheme with the constant level in the adjacent horizon [21] on the example of cluster pumping from the pressure Upper Cretaceous (Mynkuduk) aquifer within the Suzak artesian basin [27].

The Mynkuduk horizon represents here the lowest part of the section of the Upper Cretaceous aquifer complex. The horizon is underlain by Palaeozoic poorly lithified silty-argillaceous sediments, which act as a regional confining layer. The water-bearing rocks of the Mynkuduk horizon are fine and medium grained sands with gravel and a low content of clay particles.

There are almost no persistent and extended regional confining layers between the Mynkuduk and above lying Inkuduk horizons or low-permeable interlayers within the horizon. All existing ones are lenses of various areas with a thickness ranging from 1–2 to 5–10 m. Low-permeable rocks are represented by clays, silts and clay sands. As a result, the adjacent aquifers of the Upper Cretaceous complex are hydraulically connected.

The disturbing well no. 2001c and observation wells no. 2002g and no. 2003g of test cluster within the Mynkuduk horizon and at the interface with Inkuduk horizon have penetrated several rather stably thick local clay interlayers with the thickness varying from 2 to 5–7 m. These interlayers determined the peculiarities of GW filtration in the disturbance region, which correspond to the regularities of their movement in multilayer aquifers in the presence of flow through the low-permeable separating sediments.

Pumping was conducted with a submersible pump at a constant average flow rate of Q = 11.10 m^3/h (266.4 m^3/day). The flow rate over the course of the experiment was measured by the volumetric method, that is, by the duration of the filling of the measuring tank with a volume of V = 100 dm^3. Time readings for measurements were made using stopwatch with a scale with a division value of 0.2 s.

The duration of the experimental pumping was 9 days. The depth down to the piezometric level in the wells during pumping was measured by two-contact electric level gauges with a measuring tape length of 50 m and a scale interval of 0.001 m.

It should be noted that the use of gauges with a measuring tape, on which millimetre divisions are applied, does not, in fact, solve the problem of raising the accuracy class of measuring equipment. The experience of using such devices in field conditions shows that the accuracy of the depth measurements to the GW level, in general, is about 0.5 cm (primarily due to the delayed action of the signal system), that is, accuracy remains the same with the traditionally used measuring devices [13, 15, 28]. In addition, as shown in [13, 15], with a relative maximum permissible random error of depth measurement to the GW level in the well by electrical level gauge with a measuring tape with millimetre divisions, in 0.06% of cases, 50 m depth is measured with the maximum absolute error of $\pm$0.030 m, 40 m depth has $\pm$0.024 m error and, finally, 10 m depth has $\pm$0.006 m error. It clearly shows that the third digit after the decimal point in the depth down to GW level in well measured by such a level gauge does not imply any information and should be discarded; it is enough to have a scale with divisions of at least 0.005 m [13, 15].

During the processing of the results of experimental pumping, a number of corrections for systematic errors were introduced into the experimental data. First, according to the results of inclinometry in the wells of the test cluster, the distances between the axes of the filters of the wells are specified. Accordingly, the geometric parameters of the test cluster are summarised in **Table 1**. Secondly, corrections for changes in the piezometric level in the wells depending on the changes in barometric pressure P are introduced into the measured decreases in the observation well levels of the test cluster. For this, before the start of pumping, the standard M-67 aneroid barometer was used to measure the atmospheric pressure for 3 days with an interval of 4 hours. At the same time points, the same two-contact electrical level gauge was used to measure the depth down to the GW level in the wells of the test cluster.

Processing of the observation results revealed a close correlational linear relationship of the fluctuations of the piezometric level with the fluctuations of the atmospheric pressure; the correlation factors characterising the closeness of this relationship for well nos. 2002g and 2003g were 0.9752 and 0.9799, respectively.

Well no.	Distance between the axes of the wellheads of disturbing and observation wells *r*, m	Depth down to the middle part of filter axis, m	Shift of the middle part of filter axis relative to the wellhead, m		Actual distances from the centre of disturbance to the axes of filters of observation wells *r*, m
			X offset	*Y* offset	
2001c	0	500	−0.56	−2.20	—
2002g	24.81	512	−0.47	−1.11	25.18
2003g	125.19	512	−3.97	−4.09	127.54

Table 1.
The values of the horizontal shift of the filter axes in the wells of the test cluster.

From these equations, it follows that the depth down to the piezometric level z per change in atmospheric pressure *P* in 1 mm Hg changes by an average of 0.5 cm, and when the pressure drops, the depth z decreases and, conversely, as the pressure increases, *P* increases as well.

This correction is made further in the results of the level tracing. For this, throughout the experiment, barometric pressure was measured using the same aneroid barometer with 4 h intervals.

The temporal tracing graphs for the observation well nos. 2002g and 2003g presented in **Figure 1** clearly indicate the bends corresponding to the beginning of the GW flow from the adjacent parts of the Mynkuduk aquifer through low-permeable interlayers. Accordingly, the parameters of the tested part of the Mynkuduk aquifer were calculated only by the first asymptotic segments of the graphs corresponding to the filtration scheme in an isolated pressure horizon not limited in the plan [19, 29, 30]. The results of the parameter calculations are given in **Table 2**.

It is to be recalled that the computational formulae corresponding to the method of temporal tracing of a decrease in the water level in observation wells during

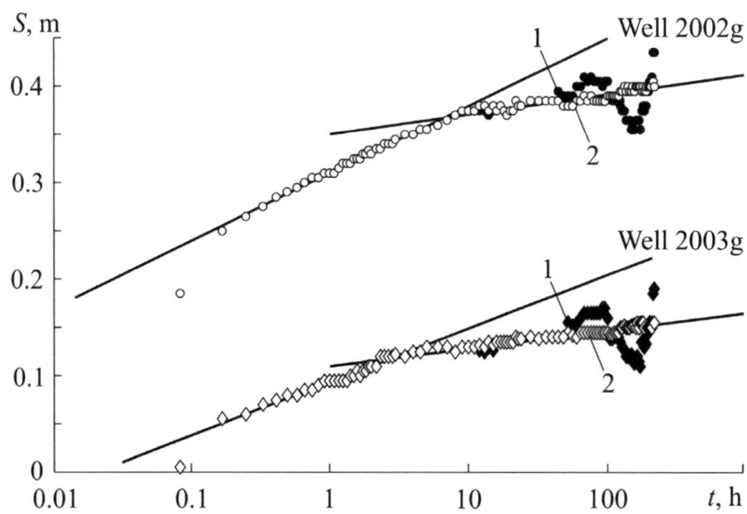

Figure 1.
Graphs of time tracking of the lowering the piezometric level in the observation wells of the experimental cluster 2002g and 2003g. Legend: 1—initial measurements of lowering the level; 2—lowering level adjusted for barometric pressure.

Parameters	Parameter values by temporal tracing data		By combined tracing data
	No. 2002g	No. 2003g	
Distance from the centre of disturbance to the axis of the filter of the observation well r, m	25.18	127.54	—
$2\sigma_r$, m	0.026	0.094	—
The number of experimental points on the calculated asymptote N	37	36	—
Slope ratio of the first calculated asymptote C, m	0.0690	0.0571	0.0684
Decrease in level A at a moment of time t_0, m	0.1890	0.0919	0.3797
Estimated pumping rate Q, m^3/h	11.10	11.10	11.10
Water transmissibility T, m^2/h	29.4	35.6	29.7
Standard error of the regression equation S_y, m	0.00242	0.00589	—
Normalised deviation of Student's distribution $t_{q,N-2}$	2.0301	2.0322	—
$2\sigma_C$, m	0.00407	0.00964	—
$2\sigma_A$, m	0.00185	0.00428	—
$2\sigma_Q$, m^3/h	0.939	0.939	—
$2\sigma_T$, m^2/h	3.04	6.72	—
Relative error of water transmissivity $2\delta_T$, %	10.33	18.87	—
B parameter value	2.731	1.610	—
Piezoconductivity χ, m^2/h	$1.55 \cdot 10^5$	$2.95 \cdot 10^5$	$1.58 \cdot 10^5$
Elastic water yield μ^*	$1.9 \cdot 10^{-4}$	$1.2 \cdot 10^{-4}$	$1.9 \cdot 10^{-4}$
$2\sigma_B$	0.1638	0.2819	—
$2\sigma_\chi$, m^2/h	$5.90 \cdot 10^4$	$1.93 \cdot 10^5$	—
Relative error of piezoconductivity $2\delta_\chi$, %	38.06	65.42	—
Relative error of elastic water yield $2\delta_{\mu*}$, %	39.44	68.01	—

Table 2.
Porosity and permeability parameters of the Mynkuduk aquifer, their random errors and errors of intermediate calculations (with a confidence probability of 0.954) established by the test cluster.

pumping, for which the solution of the problem of inflow to the well arising from the initial differential filtration equation, are approximated by the equation [19, 29, 30].

$$S = A + C\lg\ t, \tag{5}$$

and are presented in the form of [12, 21].

$$T = \frac{0.183\ Q}{C}, \quad C = \frac{S_2 - S_1}{\lg\ t_2 - \lg\ t_1}, \quad \chi = 0.445\ \frac{r^2}{t_0}\ 10^{\mathrm{B}}, \quad \mathrm{B} = \frac{A}{C}, \tag{6}$$

where C is the slope ratio of the calculated asymptotic segment in the temporal tracing graph; S_1 and S_2 are the decrease of the level in the assigned observation well at time points t_1 and t_2 from the start of pumping, respectively; A is the decrease in the level at time point t_0; r is the distance between the axes of the filters of the disturbing and the observation wells. The values S_1, S_2 and A are removed from the asymptotic part of the tracing graph, and the time t_0 is set to 1 in the selected units, so lg $t_0 = 0$.

Conversion of equations from (6) in accordance with (3) gives [12, 20].

$$\sigma_T \cong \sqrt{\left(\frac{0.183\ \sigma_Q}{S_2 - S_1} \lg \frac{t_2}{t_1}\right)^2 + \left[\frac{0.183\ Q}{(S_2 - S_1)^2} \lg \frac{t_2}{t_1}\right]^2 \left(\sigma_{S_1}^2 + \sigma_{S_2}^2\right)}, \tag{7}$$

$$\sigma_\chi \cong \sqrt{\left(0.890 \frac{r}{t_0} 10^B \sigma_r\right)^2 + \left(1.036 \frac{r^2}{t_0} 10^B \sigma_B\right)^2}, \quad \sigma_B \cong \sqrt{\left(\frac{\sigma_A}{C}\right)^2 + \left(\frac{A\,\sigma_C}{C^2}\right)^2}. \tag{8}$$

Errors in the time readings do not exceed a few seconds (in the extreme case, the first tens of seconds); therefore, when processing the experimental data, the terms of sum in (3) containing σ_t can be neglected [12, 20]; they are omitted in formulae (7) and (8).

When formulating the problem of assessing the accuracy of parameters, a priori, it was assumed that the measured decreases of the level unambiguously fall on the asymptotes of the temporal tracing graphs; this assumption is clearly manifested in Eq. (7). In this case, the fact that the experimental set of points was dispersed relative to the true asymptote, which corresponds to the actual parameters of the conducting medium, was disregarded. And yet, in a point cloud, even with a visible regression connection, in each specific case, there can be several averaging options (performing computational asymptotes), which is, among other things, the cause of individual errors. The results of the study of such individual random errors are set forth, in particular, in [23].

As is known, one of the most important representations of the macroscopic phenomenological theory of filtration is the constancy of parameters of medium over time. With reference to the graph-analytical method, this means the existence of asymptotic segments with constant slope ratio on the temporal tracing graphs. The dispersion of measured values of GW level decrease relative to these asymptotes is due to the realisation of measurement errors (instrumental error), fluctuations in the disturbance flow rate around its average value taken as the calculated one (methodical error) and the influence of chaotic geofiltration inhomogeneity of higher order or effective one (model error). Therefore, the experimentally recorded deviations of the level drops from the approximating asymptotes (which determine the final random errors of the parameters) should always exceed such deviations only due to instrumental errors.

Thus, the task of estimating the random errors of transmissibility, piezoconductivity and water yield of water-bearing sediments is reduced to find the calculated asymptotes that best match the true asymptotes and compare them with the asymptotes for the extreme variants of the drawing. Estimates of the discrepancies between them are maximal and complexly contain all possible random errors of permeability parameters [5, 12, 20].

The necessity of comparing the chosen calculated asymptote with the true one is a serious problem: within the framework of the physical and mathematical model of nonstationary GW filtration, and there are no criteria determining the preferred choice of any of the asymptotes. A very fruitful idea to get around this problem seems to be an involvement of the true (more precisely, the closest to the true) asymptotes of the regression analysis to justify this choice. Analysis of parameters A and C in (5) as regression coefficients and their confidence intervals for a selected level of significance provides the required estimate of the total random errors of the parameters. Optimization of studies of this kind is achieved on the basis of the maximum likelihood method, which, assuming a normal distribution of experimental data, is reduced to the least squares method [5, 12, 20, 21]. In [24], when justifying the method for detecting gross errors in measuring level decreases, the distribution of the deviations of these measured decreases from the calculated asymptotes justified by the least squares method was proved as complying with the normal law.

As it was noted before, the confidence intervals of the regression coefficients A and C to the selected level of significance are identified with the random errors of these coefficients to the same level of significance, and taking into account the latter, the errors of the parameters are established. Random errors A at any level of significance q were determined by the formulae [31].

$$\Delta_A = \pm \frac{t_{q,\ N-2}\ S_y}{\sqrt{N-2}} \sqrt{1 + \frac{N\left(\overline{\lg t}\right)^2}{\sum\limits_{j=1}^{N} \left(\lg\ t_j - \overline{\lg\ t}\right)^2}}, \quad S_y = \sqrt{\sum_{j=1}^{N} \left(\frac{S_j - \tilde{S}_j}{N-2.}\right)^2} \tag{9}$$

Here $\overline{\lg\ t}$ is the average value of the variable lg t_j; $t_{q,\ N-2}$ is the normalised deviation of the Student's distribution that depends on the level of significance q and the number of degrees of freedom $N-2$, the values of the deviation can be found, for example, in [26]; N is the number of measured values of the level decrease on the calculated asymptote; S_y is the standard error of the regression equation; $\tilde{S}_j$ is the value of variable S, calculated by the regression equation.

The calculation of the errors of the slope ratio C is performed on the basis of the equation for the limits of its confidence interval [31], in which the value of the confidence interval at any level of significance is as follows:

$$\Delta_C = \pm \frac{t_{q,\ N-2}\ S_y}{\sqrt{N-2}} \sqrt{\frac{N}{\sum\limits_{j=1}^{N} \left(\lg\ t_j - \overline{\lg\ t}\right)^2}}. \tag{10}$$

The method of processing experimental data provides for a constant pumping rate throughout its duration. However, in fact, its value fluctuates around a certain average value taken as a calculated one. In the works [9, 13, 15], information is given on the instrumental errors of pumping flow rate measurement by methods used in hydrogeological practice. In particular, devices for establishing flow rate by the volumetric method provide measurements with limiting relative errors of 1–6%.

In fact, the total random errors in measuring the flow rate are noticeably bigger than instrumental ones, since the recorded fluctuations Q are also due to the unstable in time operation of the water-lifting equipment. Due to the equal probability of deviation in one direction or another from the average measured values of the flow rate, their distribution can be considered normal, and then the standard quadratic deviation Q is equal to

$$\sigma_Q = \sqrt{\frac{\sum\limits_{j=1}^{N} \left(Q_j - Q\right)^2}{N-1}}. \tag{11}$$

Here, Q_j and Q are the measured and average values of flow rate, respectively.

Eq. (7) can be changed taking into account the second in (6), (10) and (11) as follows:

$$\sigma_T \cong \sqrt{\left(\frac{0.183\ \sigma_Q}{C}\right)^2 + \left(\frac{0.183\ Q\ \sigma_C}{C^2}\right)^2}. \tag{12}$$

The absolute random error of the piezoelectric conductivity is still estimated by the ratio (8).

Now, there are dependencies necessary for calculating all the errors of primary measurements and intermediate calculations for calculating random errors of permeability parameters. The results of calculations of the parameters of the Mynkuduk aquifer, their random absolute and relative errors and errors of intermediate calculations performed in accordance with the methods developed in [5, 12, 20–22] shown herein, are summarised in **Table 2**.

According to these results, the following remarks should be made. Random errors of transmissibility and piezoconductivity, established according to the level tracing data in well no. 2003g, significantly (approximately twice) exceed those for the parameters of well no. 2002g. This fact has a simple and quite logical explanation. With a sufficiently large depth down to the piezometric level in the observation wells (~28–30 m) and the same degree of dispersion of experimental points relative to the calculated asymptotes on the level tracing graphs in both wells of the test cluster, the error of the parameters is greater than for the well for which lower absolute values of level decrease are recorded. This is well no. 2003g, located at a greater distance from the centre of the disturbance.

The logarithmic approximation of the Theis formula as (5) and (6) holds for the conditions of a quasi-stationary filtration flow [19]. There is an analytical criterion (control time) t_K used to find the plot of the graph that meets the quasi-stationary mode:

$$t_K = \frac{r^2}{0.4\chi}. \tag{13}$$

All designations here remain the same.

The control time of the onset of the quasi-stationary filtration mode, the assessment of which is made on the basis of the parameters presented in **Tables 1** and **2**, in accordance with (13) is as follows: $t_K \cong 0.01$ h (36 s) for well no. 2002g, the filter of which is located at a distance of r_1 = 25.18 m from the axis of the filter of the disturbing well; $t_K \cong 0.26$ h (15 min) for well no. 2003g, the filter of which is located at a distance of r_2 = 127.54 m from the axis of the filter of the disturbing well.

The calculated values of the control time clearly showed that the selected rectilinear asymptotic segments on the temporal tracing graphs in the observation wells of the test cluster fully satisfy the applicability condition of a logarithmic approximation of the Theis formula for processing and interpreting experimental data. Accordingly, the parameters of the test aquifer, calculated for these sections, are representative.

A kind of quality control performed by the EFT for the layered aquifer system under conditions of GW flow into the test horizon from the adjacent one through separating low-permeable formation provides a combined way of processing the tracing data of the piezometric level decrease in the observation wells of the test cluster. A characteristic feature of indicator graphs of combined level tracing, as well as temporal tracing graphs, is the presence of bends that limit rectilinear asymptotic segments with different rates of change in the rate of level decrease (**Figure 2**). At the same time, the second asymptotic segments of the observation wells, which are located at different distances from the disturbance centre, deviate from the first asymptote common to the same observation wells. Thus, the match of the first asymptotes, their confluence, is an indicator of the quality of the EFT of the Mynkuduk aquifer under conditions of GW flow.

The results of calculations of the parameters of the tested Mynkuduk aquifer according to the interpretation of the combined tracing graphs are presented in **Table 2**. These parameters are in excellent agreement with the parameters established according to the interpretation of temporal tracing of the level.

2.4 Estimates of random error parameters of groundwater overflow

The results of the test pumping ensured the determination of the parameters of the GW flow from the adjacent aquifer to the tested one through low-permeable sediments. Their assessment was carried out on the basis of the filtration scheme in a multi-layered formation with a constant level in the feeding aquifer [1–4, 19, 32].

As it was noted before, the temporal tracing graphs of the piezometric level in the observation wells Nos. 2002g and 2003g, two asymptotic segments are clearly distinguished. According to the first, the parameters of the tested Mynkuduk aquifer are estimated (see **Table 2**). The second, final ones ensured the calculation of the flow factor B and the permeability factor of the separating low-permeable layer.

The inversed problem that is the estimation of $W(z)$ function is solved by the known values of Q, S_K (where S_K is the estimated level decrease in the observation well at the selected time) and T:

$$W\left(u,\ \frac{r}{B}\right) = \frac{4\pi T S_K}{Q}. \tag{14}$$

Then, using the established values of $W(z)$ and u from the tables of the Hantush function given, for example, in [3], $\frac{r}{B}$ ratio value is selected, from which the flow factor B is calculated. Taking into account the values of the latter and if the layer thickness of low-permeable sediments m_0 is known, the value of their permeability factor k_z is estimated using the formula [1–3, 19] as follows:

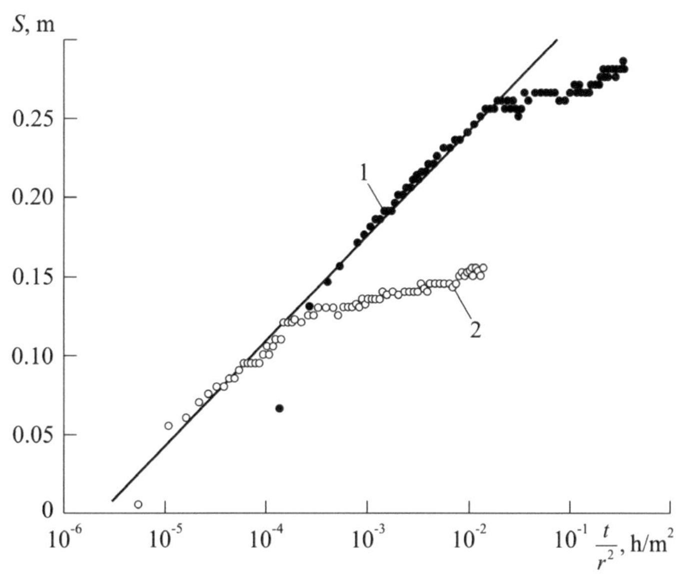

Figure 2.
Graphs of the combined tracking of the lowering the piezometric level in the observation wells of the experimental cluster 2002g and 2003g, taking into account the corrections introduced. Legend: 1—lowering the level in the well no. 2002g; 2 – lowering the level in the well no. 2003g.

$$k_z = \frac{T m_0}{B^2}. \tag{15}$$

The results of the calculations of the flow factor and the permeability factor of low-permeable sediments in the observation wells of the test cluster are summarised in **Table 3** [21]. The values of the flow factor and permeability factor of the separating layer, averaged over the observation wells of the test cluster, are $B \cong 1810$ m and $k_z \cong 7.1 \cdot 10^{-5}$ m/h.

Applying formula (3) to Eq. (14) gives the following equation for the root mean square deviation of the function $W(z)$ [21]:

Parameters	**Parameter values by observation wells of the cluster**	
	No. 2002g	**No. 2003g**
Average pumping rate Q, m^3/h	11.10	11.10
Distance from the centre of disturbance to the axis of the filter of the observation well r, m	25.18	127.54
Estimated moment of time from the start of pumping t, h	20	20
Estimated level decrease S_K, m	0.257	0.135
Number of experimental points on the calculated (second) asymptote N	63	65
Standard error of the regression equation S_y, m	0.003862	0.002150
Normalised deviation of Student's distribution $t_{t,\ N-2}$	1.9994	1.9986
$2\ \sigma_{S_K}$, m	0.003254	0.001668
Value of Hantush function $W(z)$	8.554	5.441
$2\ \sigma_{W(z)}$	1.148	1.127
$2\ \delta_{W(z)}$, %	13.42	20.71
Estimated value of u parameter	$5.11 \cdot 10^{-5}$	$6.89 \cdot 10^{-4}$
$2\ \sigma_u$	$1.95 \cdot 10^{-5}$	$4.52 \cdot 10^{-4}$
$2\ \delta_u$, %	38.16	65.60
$\frac{r}{B}$ ratio	0.0137	0.0722
Relative error $2\delta_{\frac{r}{B}}$, %	40.45	68.79
$2\sigma_{\frac{r}{B}}$	0.0055	0.0497
Value of flow factor B, m	1840	1770
$2\sigma_B$, m	740	1220
$2\delta_B$, %	40.22	68.93
Thickness of the layer of low-permeable sediments m_0, m	7.0	7.0
$2\sigma_{m_0}$, m	0.50	0.50
Permeability factor for the separating layer k_z, m/h	$6.1 \cdot 10^{-5}$	$8.0 \cdot 10^{-5}$
$2\sigma_{k_z}$, m/h	$4.9 \cdot 10^{-5}$	$1.1 \cdot 10^{-5}$
$2\delta_{k_z}$, %	80.33	137.50

Table 3.
Values of flow factor and porosity and permeability parameters of low-permeable sediments and their random errors (with a confidence probability of 0.954).

$$\sigma_{W(z)} \cong \sqrt{\left(\frac{4\pi S_K \sigma_T}{Q}\right)^2 + \left(\frac{4\pi T \sigma_{S_K}}{Q}\right)^2 + \left(\frac{4\pi T S_K \sigma_Q}{Q^2}\right)^2}. \tag{16}$$

As it was mentioned before, the error of the calculated decrease taken from the asymptote for any lg t_K at any selected time point with a confidence level of 0.683 is expressed by the following equation:

$$\sigma_{S_K} = \pm \frac{t_{q,\ N-2}\ S_y}{\sqrt{N-2}} \sqrt{1 + \frac{N\ \left(\lg\ t_K - \overline{\lg\ t}\right)^2}{\sum_{j=1}^{N}\left(\lg\ t_j - \overline{\lg\ t}\right)^2}}. \tag{17}$$

Eq. (17) is the basis of experimental estimates of σ_{S_K}.

So, now there is a possibility of finding random errors of the values of the Hantush function (used later to determine the flow factor B and the permeability factor of low-permeable sediments k_z), and, consequently, the errors of the flow parameters (see **Table 3**).

A few comments should be made concerning the order of calculations.

The calculation of the random errors of $\frac{r}{B}$ parameter based on the original equation of the Hantush function from [3] is impossible in principle. To implement this procedure, the following technique was proposed [21].

Random errors of the Hantush function are quite simply calculated from Eq. (16). The errors of the parameter u, one of the arguments of the Hantush function, are estimated from the following equation (neglecting the term of sum containing σ_t).

$$\sigma_u \cong \sqrt{\left(\frac{r\sigma_r}{2\chi t}\right)^2 + \left(\frac{r^2\sigma_\chi}{4\chi^2 t}\right)^2},$$

where all the designations remain the same.

The ratio $\frac{r}{B}$, as provided by the method [19], is calculated using the Hantush function tables based on the following equation:

$$\frac{r}{B} = Inf\, W(z). \tag{18}$$

In the presence of the calculated absolute values of functions $W(z)$ and u, as well as their random errors $\sigma_{W(z)}$ and σ_u, it is easy to determine the relative errors of functions $\delta_{W(z)} = \frac{\sigma_{W(z)}}{W(z)}$ and $\delta_u = \frac{\sigma_u}{u}$ for the same level of confidence probability. Accordingly, regardless of the type of function $W(z)$ in (18), the relative random error of the ratio $\frac{r}{B}$ can be represented as follows:

$$\delta_{\frac{r}{B}} \cong \sqrt{\delta^2_{W(z)} + \delta^2_u},$$

whence it is easy to estimate $\sigma_{\frac{r}{B}}$.

Having denoted the calculated value of ratio $\frac{r}{B}$ as G, and assigning the random error calculated above, that is, $\sigma_{\frac{r}{B}} = \sigma_G$ to the last parameter, the equation for the random error B can be presented as follows:

$$\sigma_B \cong \sqrt{\left(\frac{\sigma_r}{G}\right)^2 + \left(\frac{r\sigma_G}{G^2}\right)^2}. \tag{19}$$

The calculation of the random error σ_B by formula (19) is provided with all the necessary intermediate values (**Table 3**) and does not cause difficulties.

Based on the ratio (15), it is possible to immediately calculate the random errors of the permeability factor of low-permeable sediments. If formula (3) is applied to Eq. (15), then [21]

$$\sigma_{k_z} \cong \sqrt{\left(\frac{m_0\,\sigma_T}{B^2}\right)^2 + \left(\frac{T\,\sigma_{m_0}}{B^2}\right)^2 + \left(\frac{2T\,m_0\,\sigma_B}{B^3}\right)^2}. \qquad (20)$$

In the latter ratio, only the value of the error in measuring the thickness of the layer of low-permeable sediments σ_{m_0} remains undetermined. If assumed that the thickness of the layer is calculated as the difference between two measurements of the depth to its bottom z_{Π} and its top z_K (for example, drill stems and geophysical instrument cable) with a measuring tape that provides the maximum permissible random error of 0.1% with the rounding of the measurement result to 1 cm, then.

$$m_0 = z_{\Pi} - z_K.$$

Then the random instrumental error of the layer thickness with a confidence level of 0.683 will be as follows [21]:

$$\sigma_{m_0} \cong \sqrt{\left(\frac{1}{3} z_{\Pi}\,0.001 + 0.01\right)^2 + \left(\frac{1}{3}\ z_K\,0.001 + 0.01\right)^2}. \qquad (21)$$

If a layer of a relatively small thickness (~5–8 m) is located at a depth of about 500 m, then $2\sigma_{m_0} \cong 0.50$ m.

2.5 Estimates of systematic errors of parameters of the tested aquifer and separating low-permeable sediments

The basis of the above methods for calculating the parameters of the tested aquifer and separating low-permeable sediments is a rather artificial assumption that the characteristics of the first asymptotic segments of the temporal and combined level tracing graphs do not show the GW flow from the adjacent horizon; therefore, this flow does not affect the values of the determined parameters, test aquifer and separating low-permeability sediments. In fact, the flow begins to manifest itself almost immediately with the start of pumping. Therefore, in fact, certain parameters contain systematic errors significant in absolute values [33, 34].

To determine the parameters from the values of which such systematic errors are excluded, original methods were proposed in [33, 34]. Here, the probable values of the indicated systematic errors established by comparing the calculated parameters of the tested Mynkuduk aquifer and separating low-permeable layer obtained using one of these methods from [33, 34] are considered. The method uses the ratio of the piezometric level decreases in combination with the method of selection given in **Tables 1** and **2**.

In accordance with this method, the primary processing of temporal tracing graphs in the observation wells of the test cluster is performed in the same way as recommended in [5, 12, 20, 21]. As a result of this processing, the logarithmic trends of the second asymptotic segments of the level tracing graphs in the observation wells are established. Accordingly, in the subsequent processing of the experimental data, either the calculated values of the decrease in the piezometric level at the designated moments of time, calculated on the basis of the equations of these trends, or the values of the decrease, recorded directly from the second asymptotic segments, are used.

Sternberg [35] obtained an alternative representation of Eq. (3) [32] as follows:

$$S = \frac{Q}{2\pi T} K_0(z) = \frac{Q}{2\pi T} K_0\left(.r \sqrt{\frac{1}{2\chi t} + \frac{1}{B^2}}\right). \tag{22}$$

Here $K_0(z)$ is the modified Bessel function of the second type of zero order. The rest designations remain the same.

Having presented Eq. (22) for $S^{(1)}$ and $S^{(2)}$, which are the decreases in GW level in observation wells as applied to the second asymptotic segments of the tracing graphs in the observation wells of the test cluster located at the distances r_1 and r_2 from the disturbance centre at the selected time point t_i, and having taken their ratio after simple transformations (provided that the tested aquifer is assumed to be homogeneous and isotropic, that is, $T = const$), the following is obtained [33, 34]:

$$\frac{S^{(1)}}{S^{(2)}} = \frac{K_0\left(r_1 \sqrt{\frac{1}{2\chi t_i} + \frac{1}{B^2}}\right)}{K_0\left(r_2 \sqrt{\frac{1}{2\chi t_i} + \frac{1}{B^2}}\right)}, \tag{23}$$

The rest designations remain the same.

Based on (23) in [33, 34], an algorithm was developed for calculating the water transmissibility T of the test aquifer, the piezoconductivity χ and the flow factor B, implemented on the basis of the probabilities in the MS Excel spreadsheet.

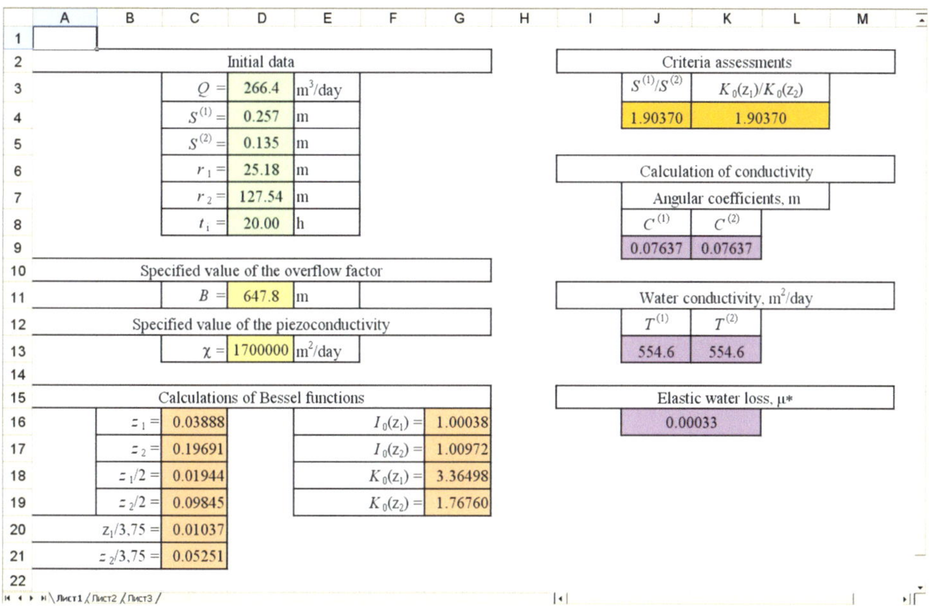

Figure 3.
Spreadsheet page with the results of the processing of groundwater level tracking data in observation wells of the pilot cluster using the drop ratio method.

Moreover, in [33, 34], the calculation of the Bessel function $K_0(z)$ was performed using its approximation by polynomials [36].

Figure 3 shows a spreadsheet page with the calculation of the flow factor *B*, water transmissibility *T* and elastic water yield μ^* from observation wells of the test cluster. The initial data on the piezometric level decreases were recorded by the second asymptotic segments of the temporal tracing graphs, and the value of $\chi = 1.7{\cdot}10^6$ m^2/day was used as the piezoconductivity, close to that calculated by the results of processing and interpreting the first asymptotic segments of the tracing graphs for the GW level decrease in the observation wells of the test cluster.

The selected value of the flow factor *B* is equal to $B = 647.8$ m, and the calculated values of the water transmissibility *T* and the elastic water yield μ^* are $T = 554.6$ m^2/day and $\mu^* \cong 3.3{\cdot}10^{-4}$, respectively (**Figure 3**). If the thickness of the separating layer $m_0 = 7.0$ m [21] is calculated by formula (15), then the value of its permeability factor is $k_z = 9.25{\cdot}10^{-3}$ m/day (or $3.85{\cdot}10^{-4}$ m/h).

3. Discussion of the obtained results

The relative random errors of water transmissibility *T* and piezoconductivity χ determined by the results of temporal level tracing in the observation wells of the test cluster and equal, respectively, to 10.33–18.87% and 38.06–65.42% with a confidence probability of 0.954, generally correspond to the level of such errors of these parameters determined from the results of EFT in aquifers [12, 20]. At the same time, the random errors of water transmissibility and piezoconductivity, established by the data of level tracing in well no. 2003g, significantly (approximately twice) exceed those for the parameters of well no. 2002g. This fact has a simple and quite logical explanation. With a sufficiently large depth down to the piezometric level in the observation wells ($\sim$28–30 m) and the same degree of dispersion of experimental points relative to the calculated asymptotes on the level tracing graphs in both wells of the test cluster, the error of the parameters is greater than for the well for which lower absolute values of level decrease are recorded. This is well no. 2003g, located at a greater distance from the centre of the disturbance.

Random errors of water transmissibility and especially piezoconductivity very significantly affect the reliability of the determined parameters of the flow through the thickness of low-permeable sediments and represent flow factor *B* and the permeability factor k_z. The relative random errors of the first of these parameters for the observation wells of the test cluster are 40.22–69.93%, that is, the values of the parameters set are certainly significant for both observation wells, although they are of unequal accuracy. The second parameter for well no. 2002g, which is closest to the centre of the disturbance, is also significant; its relative random errors are about 80.33%. For well no. 2003g, which is located at a greater distance from the centre of disturbance, these parameters are not significant at first glance, since their relative random errors exceed the established values of these parameters and represent about 137.50%. The reason for this, as in the case of piezoconductivity, seems to be the smaller absolute value of the level decrease in well no. 2003g, than in well no. 2002g.

The application of the method of the ratio of the GW level decrease in the observation wells of the test cluster ensured the identification of systematic errors in the parameters of the tested aquifer and the parameters of the separating low-permeable formation determined by traditional methods. Thus, the value of the water transmissibility T of the test aquifer obtained by the method of the level decrease ratio is T = 554.6 m^2/day and is significantly differs by 27.23–54.06% from the values determined according to the temporal and combined level tracing in the observation wells of the test cluster [21] (respectively, $T_{2002g} = 705.6$ and $T_{2003g} = 854.4$ m^2/day for temporal tracing, and T = 712.8 m^2/day for combined tracing). Even more significant are the differences in the values of the flow factor established by different methods; the following values are given in [21]: B_{2002g} = 1840 m and B_{2002g} = 1770 m. These values are 2.84 times and 2.73 times greater than the values determined by the method of level decrease ratio, respectively.

4. Conclusion

As in the case of a two-layered formation [21, 22], a paradoxical situation is created—the model correctly represents and describes the actual process of GW filtration in a layered formation, and the effective parameters of low-permeable separating sediments included in this model are set with errors exceeding their absolute values. Nevertheless, as in the case of two-layered formation, such parameters should be recognised as significant. With relative random errors exceeding 100%, the parameters of low-permeable sediments cannot take on a zero or negative value, since this contradicts the original physical and mathematical model. Accordingly, the errors should be presented only in the form of relative, expressed in multiplicity values relative to the most probable values of the set parameters.

The application of the method of the ratio of the GW level decrease MW in the observation wells of the test cluster ensured the identification of systematic errors in the parameters of the tested aquifer and the parameters of the separating low-permeable formation. It is clear that as a calculation and the most likely parameters, the parameters determined by the method of the ratio of level decrease should be used. In this case, the above random errors of these parameters should be used as estimates of the random errors of the parameters, determined, among other things, by the method of the ratio of level decreases.

Such peculiarities of the ratio between the parameters of the tested aquifers and the separating low-permeable sediments in layered aquifer systems established according to the results of their EFT and their random errors appear to be characteristic of all types of the structure of these systems.

Author details

Mikhail M. Burakov
Limited Partnership NPPF "KazGIDEK", Almaty, Kazakhstan

*Address all correspondence to: michael.burakov@gmail.com

IntechOpen

© 2019 The Author(s). Licensee IntechOpen. This chapter is distributed under the terms of the Creative Commons Attribution License (http://creativecommons.org/licenses/by/3.0), which permits unrestricted use, distribution, and reproduction in any medium, provided the original work is properly cited.

References

[1] Hantush MS. Modification of the theory of leaky aquifers. Journal of Geophysical Research. 1960;**65**: 3713-3725

[2] Hantush MS, Jacob CE. Nonsteady radial flow in an infinite leaky aquifer. Transactions of the American Geophysical Union. 1955;**36**:95-100

[3] Hantush MS. Analysis of data from pumping tests in leaky aquifers. Transactions of the American Geophysical Union. 1956;**37**:702-714

[4] Plugina TA. Opredeleniye Geofil'tratsionnykh Parametrov Slabopronitsayemykh Otlozheniy Naturnymi Metodami (Determination of Geofiltration Parameters of Low-Permeable Sediments Using Natural Methods). Obzor. VIEMS. Gidrogeologiya i Inzhenernaya Geologiya. Moscow: VIEMS. 1978. p. 56. (in Russian)

[5] Burakov MM. Metodicheskiye Osnovy Otsenok Sluchaynykh Pogreshnostey Fil'tratsionnykh i Yemkostnykh Parametrov Vodonosnykh Gorizontov (Methodical Bases for Estimating Random Errors of Filtration and Capacitive Parameters of Aquifers). Almaty: ProService LTD; 2007. p. 74. (in Russian)

[6] Zheleznyakov GV, Danilevich BB. Tochnost' Gidrologicheskikh Izmerenii i Raschetov (Accuracy of Hydrological Measurements). Leningrad: Gidrometeoizdat; 1966. p. 240. (in Russian)

[7] Malikov SF, Tyurin NI. Vvedenie v Metrologiyu (Introduction to Metrology). Standartgiz: Moscow; 1966. p. 248. (in Russian)

[8] Mudrov VI, Kushko VL. Metody Obrabotki Izmerenii. Kvazipravdopodobnye Otsenki (Measurement Processing Data. Quasi-Plausible Estimates). Moscow: Sovetskoye Radio; 1976. p. 192. (in Russian)

[9] Il'in NI, Chernyshev SN, Dzektser ES, et al. Otsenka Tochnosti Opredeleniya Vodopronitsaemosti Gornykh Porod (Assessment of the Accuracy of Rock Permeability). Moscow: Nauka; 1971. p. 151. (in Russian)

[10] Sveshnikov AA. Osnovy Teorii Oshibok (Fundamentals of Error Theory). Izd. LGU: Leningrad; 1972. p. 125. (in Russian)

[11] Shenk K. Teoriya Inzhenernogo Eksperimenta (Theory of Engineering Experiment). Moscow: Mir; 1972. p. 381. (in Russian)

[12] Burakov MM. On the reliability of assessments of aquifer parameters. Water Resources. 1996;**23**(5):539-547

[13] Dubinchuk VT, Shustov VM. Izmereniya Pri Gidrogeologicheskikh i Inzhenerno-Geologicheskikh Issledovaniyakh (Measurements in Hydrogeological and Geotechnological Studies). Vol. 192. Moscow: Nedra; 1984. (in Russian)

[14] Pertsovskii VV. Ob otsenke tochnosti opredeleniya parametrov fil'tratsii (assessing the determination accuracy of hydraulic characteristics). Razvedka i Okhrana Nedr. 1968;**4**: 41-43. (in Russian)

[15] Shustov VM. Tekhnika Izmerenii Pri Polevykh Gidrogeologicheskikh Issledovaniyakh (Measurement Techniques in Field Hydrogeological Studies). Moscow: Nedra; 1978. p. 193. (in Russian)

[16] Bindeman NN. Otsenka Ekspluatatsionnykh Zapasov

Podzemnykh Vod. Metodicheskoye Rukovodstvo (Evaluation of Groundwater Reserves. Methodological Manual). Moscow: Gosgeoltekhizdat; 1963. p. 204. (in Russian)

[17] Bindeman NN, Yazvin LS. Otsenka Ekspluatatsionnykh Zapasov Podzemnykh Vod. Metodicheskoye Rukovodstvo (Evaluation of Groundwater Reserves. Methodological Manual). Moscow: Nedra; 1970. p. 216. (in Russian)

[18] Bochever FM, Garmonov IV, Lebedev AV, Shestakov VM. Osnovy Gidrogeologicheskikh Raschetov (Basics of Hydrogeological Calculations). Moscow: Nedra; 1965. p. 307. (in Russian)

[19] Borevskii BV, Samsonov BG, Yazvin LS. Metodika Opredeleniya Parametrov Vodonosnykh Gorizontov po Dannym Otkachek (Methods for Determining Aquifer Characteristics by Pumping Test Data). Moscow: Nedra; 1979. p. 326. (in Russian)

[20] Burakov MM. Ob otsenkakh dostovernosti fil'tratsionnykh i yemkostnykh parametrov, poluchennykh po dannym proslezhivaniya vosstanovleniya urovnya podzemnykh vod (on the estimates of reliability of hydraulic and storage characteristics derived from the data on groundwater level recovery). Gidrometeorologiya i Ekologiya. 2005;**2**: 153-170. (in Russian)

[21] Burakov MM. Sluchainye Pogreshnosti Parametrov Sloistykh Vodonosnykh Sistem, Ustanovlennykh po Dannym Otkachek (Random Errors in Multi-Layer Aquifer Characteristics Derived from Pumping Tests). ProService LTD: Almaty; 2007. p. 113. (in Russian)

[22] Burakov MM. Random errors in the hydrogeological parameters of a two-layer stratum evaluated by multi-well pumping tests. Water Resources. 2017; **44**(4):614-625

[23] Burakov MM. Lichnyye sluchaynyye pogreshnosti fil'tratsionnykh parametrov (personal random errors of filtration parameters). Izvestiya VUZov, Geologiya i Razvedka. 2000;**6**:103-115. (in Russian)

[24] Burakov MM. Metody analiza izmerennykh znacheniy ponizheniya urovnya pri opytnykh otkachkakh na nalichiye grubykh oshibok (methods for analyzing the measured values of the decrease in the level with experimental pumping for the presence of gross errors). Izvestiya VUZov, Geologiya i Razvedka. 2014;**5**:43-53. (in Russian)

[25] Demidovich BP, Maron IA. Osnovy Vychislitel'noi Matematiki (Fundamentals of Computational Mathematics). Moscow: Nauka; 1970. p. 460. (in Russian)

[26] Bol'shev LN, Smirnov NV. Tablitsy Matematicheskoi Statistiki (Tables of Mathematical Statistics). Moscow: Nauka; 1983. p. 416. (in Russian)

[27] Dmitrovsky VI, editor. Gidrogeologiya SSSR. T. XXXVI. Yuzhnyy Kazakhstan (Hydrogeology of the USSR. T. XXXVI. South Kazakhstan). Moscow: Nedra; 1970. p. 460. (in Russian)

[28] Shestakov VM, Bashkatov DM, Pashkovsky IS, et al. Opytno-fil'tratsionnyye Raboty (Experimental Filtration Works). Moscow: Nedra; 1974. p. 203. (in Russian)

[29] Mironenko VA, Shestakov VM. Teoriya i Metody Interpretatsii Opytno-fil'tratsionnykh Rabot (Theory and Methods of Pumping Test Interpretation). Moscow: Nedra; 1978. p. 326. (in Russian)

[30] Gavich IK, Kovalevsky VS, Yazvin LS, et al. Osnovy Gidrogeologii.

Gidrogeodinamika (Basics of Hydrogeology. Hydrogeodynamics). Novosibirsk: Nauka; 1983. p. 241. (in Russian)

[31] Seber J. Lineynyy Regressionnyy Analiz (Linear Regression Analysis). Moscow: Mir; 1980. p. 456. (in Russian)

[32] Sindalovsky LN. Spravochnik Analiticheskikh Resheniy Dlya Interpretatsii Opytno-fil'tratsionnykh Oprobovaniy (Handbook of Analytical Solutions for the Interpretation of Test Filtration Tests). SPb: Izd. S.-Peterb. Universiteta; 2006. p. 796. (in Russian)

[33] Burakov MM. Opytnyye Otkachki Iz Sloistykh Vodonosnykh Sistem s Peretekaniyem. Metody Interpretatsii Rezul'tatov (Experienced Pumping from Layered Aquifer Systems with Overflow. Methods of Interpretation of the Results). Saarbrücken: LAP LAMBERT Academic Publishing; 2014. p. 145. (in Russian)

[34] Burakov MM, Khabiev SK. Metodika Opredeleniya fil'tratsionnykh i Yemkostnykh Parametrov Sloistykh Vodonosnykh Sistem po Dannym Kustovykh Otkachek (the Method for Determining the Filtration and Capacitive Parameters of Layered Aquifer Systems According to the Pumping Data). Almaty: AO "NTSNTI"; 2010. p. 67. (in Russian)

[35] Sternberg YM. Some approximate solutions of radial flow problems. Journal of Hydrology. 1969;**33**(2): 158-166

[36] Abramowitz M, Stegun I, editors. Spravochnik po Spetsial'nym Funktsiyam s Formulami, Grafikami i Matematicheskimi Tablitsami (Handbook of Special Functions with Formulas, Graphs and Mathematical Tables). Moscow: Nauka; 1979. p. 831. (in Russian)

Chapter 4

Positive Effect of Climate Change on Water Resources Enhancement in Africa: Case of Gambia River Basin (Senegal)

Cheikh Faye

Abstract

With climate change and the return of wet years over the last few years, West African rivers have begun to record excess flows and rising levels. These new changes have resulted in a slight increase in water resources. The objective of this study is therefore to assess the climate change impacts on water resources in the Gambia River. The methodology is based on the calculation of normalized indices (rainfall and flow) and the application of the Maillet model. The results show that just as rainfall increased on both sides of 1994 at the Kédougou (16.1%) and Tambacounda (13.3%) stations, flows in the Gambia basin rose which can sometimes exceed 50% at certain stations (34.7% in Simenti and so on). In addition, the depletion coefficients, on both sides of 1994, vary between 0.051 and 0.044 at the Mako station (a decrease of −15.1%) and between 0.057 and 0.045 at Simenti (a decrease of −21.3%). These coefficients highlight an increase in the duration of drying up and an increase in the volumes of water mobilized by the aquifers that go from 3035 to 6531 m^3 in Mako (an increase of 115%) and from 3017 to 4581 m^3 in Simenti (an increase of 17%).

Keywords: climate change, water resources, standardized indices, statistical tests, Gambia River basin

1. Introduction

Climate change indicates the trend of global and multi-year increases in mean sea-temperature and atmospheric temperature, the decrease in rainfall and quantity of rainfall in some regions, and increased natural disasters and extreme events (drought, floods, etc.) [1] Global warming combined with the increased variability of rainfall is leading to an upsurge in extreme events, including floods and low flows, which will increase in frequency and in intensity across the African continent Various studies highlight the evolution of flows in rivers and the impacts on the natural and human systems of the territories, during the recent period The extent and reality of this evolution depend on the regions and hydro-climatic conditions [2].

Africa is at the forefront of the regions concerned by the impact of climate fluctuations on water resources [3]. Several studies carried out in West and Central

Africa have shown, from the 1970s, a decrease in surface and groundwater flow as a result of the decrease in rainfall [4–10], and from the 1990s an increase in flow [11–17], foreshadowing an improvement of the hydrological regime in this space.

Over the whole of the Gambia watershed, prolonged conditions of rainfall and hydrometric deficits since the 1970s have been highlighted [4, 7]. The objective of this study is to assess the impacts of climate change on surface and groundwater resources in the Gambia watershed. This study aims to provide decision-makers with the necessary elements for the implementation of water resource management policies adapted to the climatic context and sustainable development of the watershed of the Gambia watershed. This study is based on the assumption that the high climatic variability observed during these last decades would have led to a decrease and a slight increase of the water resources in succession to the years 1970 and 1990. The methodology was based on the indices and the Maillet model and consisted of analyzing rainfall and runoff series variations on both sides of 1994, which is the year of rupture of rainfall and hydrological data from the Gambia watershed [7].

2. Study area

The Gambia Basin covers an area of nearly 77,100 km, shared between three states [18]: Guinea (where it originates from 1125 m of attitude around Labé); Senegal (from which it drains almost the entire Tambacounda region and part of Upper Casamance and South Sine-Saloum); Gambia (of which he is the backbone). It extends, in latitude, from 11° 22 North (in the Fouta-Djalon) to 14° 40 North (in the Far-Eastern Ferlo) and, in longitude, from 11° 13 West (Fouta-Djalon) to the 16 ″42 West (Banjul, mouth), which covers the territories of Guinea, Senegal and The Gambia, and is divided into two distinct zones: the continental basin and the estuary basin, respectively to the east and south. West of Gouloumbo, the last station where the flow of freshwater inputs is measured [4, 7]. The 1180 km long river Gambia has its source at about 1150 m altitude in the Fouta Djalon, near Labé

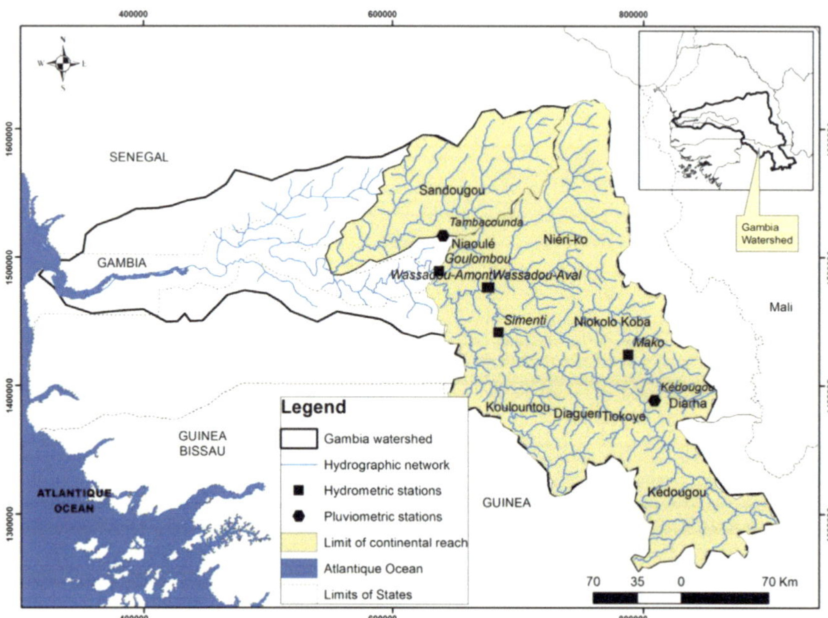

Figure 1.
The Gambia River basin situation.

Three main areas of relief can be distinguished: the maritime part, the central part and the high-basin with very marked relief The Gambia river network upstream of Gouloumbou is very dense, with many tributaries that contribute to the flow The Gambia Basin is subject to tropical climate with a long dry season e (November to May) and a short rainy season (June to October), it is mostly located in the Sudano - Guinean zone. The Gambia Basin is located in the band of isohyets 1700–700 mm. The maritime part is under the isohyet 1000 mm. North of this isohyet, contributions to the river are very low and almost negligible in the water balance (**Figure 1**).

3. Data and methods

3.1 Data

For this study, rainfall and minimum and maximum temperature data from the Tambacounda and Kedougou stations from 1950 to 2012 are used for characterization of climate change in the Gambia Basin (**Table 1**). For the impacts of climate change on surface water resources in the Gambia Basin, the hydrological stations of Gouloumbou, Mako, Simenti, Wassadou Amont and Wassadou Aval from 1953 to 2014 are used.

Finally, to characterize the impacts of climate change on groundwater resources in the Gambia Basin, the Mako station from 1970 to 2014 was used. The selected stations obey, according to the evaluated parameter, criteria of continuity, duration of the available information and data quality.

3.2 Methods

3.2.1 Calculating the standardized precipitation index (ISP) or flow rates (ISD)

The standardized precipitation index [19, 20] was developed to quantify the rainfall deficit for multiple time scales. Adopted in 2009 by the World Meteorological Organization (WMO) as a global instrument for measuring meteorological droughts, according to the "Lincoln Declaration on Drought Indices" [21], it is expressed mathematically as follows:

$$SPI\ (SFI) = \frac{(X_i - X_m)}{S}$$

Categories	Stations	Latitude (° decimals)	Longitude (° decimals)	Altitude (m)	Période
Climatology	Tambacounda	13.77	−13.68	49	1963–2012
	Kédougou	12.57	−12.22	178	1963–2012
Hydrology	Gouloumbou	13.46	−13.71		1963–2012
	Mako	12.84	−12.35		1963–2012
	Simenti	13.03	−13.3		1963–2012
	Wassadou upstream	13.34	−13.36		1963–2012
	Wassadou Downstream	13.34	−13.37		1963–2012

Table 1.
Rainfall stations in Senegal selected for the study and their characteristics.

SPI values	Drought sequences	SPI values	Wet sequences
0.00 < SPI < −0.99	Slightly dry	0.00 < SPI <0.99	Slightly wet
−1.00 < SPI < −1.49	Moderately dry	1.00 < SPI <1.49	Moderately wet
−1.50 < SPI < −1.99	Severely dry	1.50 < SPI <1.99	Severely wet
SPI < −2.00	Extremely dry	2.00 < SPI	Extremely wet

Table 2.
Classification of dryness sequences of moisture.

With Xi: the rain (or the flow) of the year i; Xm: the mean (or mean) rainfall of the series over the time scale considered; S: the standard deviation of the series on the time scale considered.

In hydrology, the standardized flow rate index (SFI) is similar to it and has been developed to quantify the water deficit for multiple time scales that will reflect the impact of drought on the availability of different types of water resources for a given period [22]. The classification of dryness sequences of moisture is given in **Table 2**.

3.2.2 Calculation of the depletion coefficient and the volume of water mobilized by aquifers

The main depletion, by the volumes that it implies and its representativity of all the aquifers of the basin, constitutes an important characteristic of the tropical hydrological regime [23]. The calculation of the depletion coefficient is based on the Maillet model [7, 9, 24]. These authors have shown the relevance of this model in the study of drying up. Maillet's model admits that in the uninfluenced regime, that is to say in the absence of any precipitation, the depletion corresponds to the exponential decay of the flow as a function of time. This is the period during which the emptying of groundwater is the only contribution to the flow of watercourses of a basin. The expression of the model of Mallet is the following:

$$Q_t = Q_0 \ e^{-\alpha(t-t_0)}$$

With Q_0 the initial flow at time t_0, $(t - t_0)$, the time expressed in days between the observation of the flow rate Q_0 and that of the flow rate Qt (flow at the end of the dry period) and α the dry-out coefficient.

The volumes of water mobilized by the aquifers are indicated calculated by the following formula:

$$V_{mobilized} = \frac{Q_0}{k}$$

The Maillet model made it possible to determine the temporal evolution of drying coefficients and water volumes mobilized by aquifers in the Gambia watershed, and to assess the duration of drying up of rivers under the effect of climate change. Thus, methods based on trend detection and break in series are applied.

3.2.3 Pettitt and Mann-Kendall tests

The Pettitt test [25] examines the existence of a break at an unknown instant in the series from a formulation derived from that of the Mann-Whitney test. This test is based on the signs of the differences in values that make up the sample.

A resulting time series is developed. The value p of the statistic makes it possible to know if this break is statistically significant at the threshold.

The Mann-Kendall test detects the presence of a linear trend (up or down) within a time series. This tendency test was first studied by Mann [26] then taken up by Kendall [27] and improved by Hirsch and Slack [28]. The robustness of the test has been validated by several comparison tests carried out by Yue and Wang [29].

4. Results

4.1 Temperature and precipitation trend

To characterize climate change in the Gambia Basin, trend shifts or breakage are sought in annual temperature and precipitation series. **Table 3** shows the results of the Pettitt and Mann-Kendall tests on these series at the Ziguinchor station over the period 1970–2012. On the minimum, maximum and average temperatures, both tests (Pettitt and Mann-Kendall) indicate the presence of a break and/or trend. This break is noted in 1987 in Kédougou and 1990 in Tambacounda for TX and TN. These breaks are confirmed by the Mann-Kendall test which shows positive and significant Kendall τ with 0.511°C for TX and 0.332°C for TN at Tambacounda. At the Kedougou station, the Kendall τ, although not significant, are positive: 0.017°C for TX and 0.726°C for TN.

To quantify the variation of the temperatures through the rupture date, we cut the time series in two subperiods on both sides of the rupture dates. The comparison of the two sub-periods shows that the one after rupture records, with respect to that before rupture, a surplus of 2.61% for TX and 1.35% for TN at Tambacounda. The non-stationarity of the temperature series from 1990 onwards is shown in **Figure 2** of the annual temperatures, which shows the average warming during the second period.

	Descriptors		Temperature			Annual precipitation
			TX	TN	TM	
Mann-Kendall test	Tambacounda	p-value	<0.0001	0.958	0.002	0.492
		τ of Kendall	0.511*	−0.006	0.332*	0.074
		Slope of Sen	0.035	0	0.014	2.158
	Kedougou	p-value	0.017	0.141	0.726	0.359
		τ of Kendall	0.259	−0.167	0.041	0.098
		Slope of Sen	0.015	−0.019	0.002	2.257
	Descriptors		**Temperature**			**Annual precipitation**
			TX	**TN**	**TM**	
Pettitt test	Tambacounda	Date of rupture	1990	1979	1990	1994
		% surplus or deficit	2.61	−0.65	1.35	13.3
	Kedougou	Date of rupture	1987	1987	1987	2003
		% surplus or deficit	−2.61	−4.27	−0.76	21.0

(−): negative trend; (+): positive trend; (*): significant trend; TX = maximum temperatures; TN = minimum temperatures; TM = average temperatures.

Table 3.
Results of the annual temperature and precipitation tests (1970–2012).

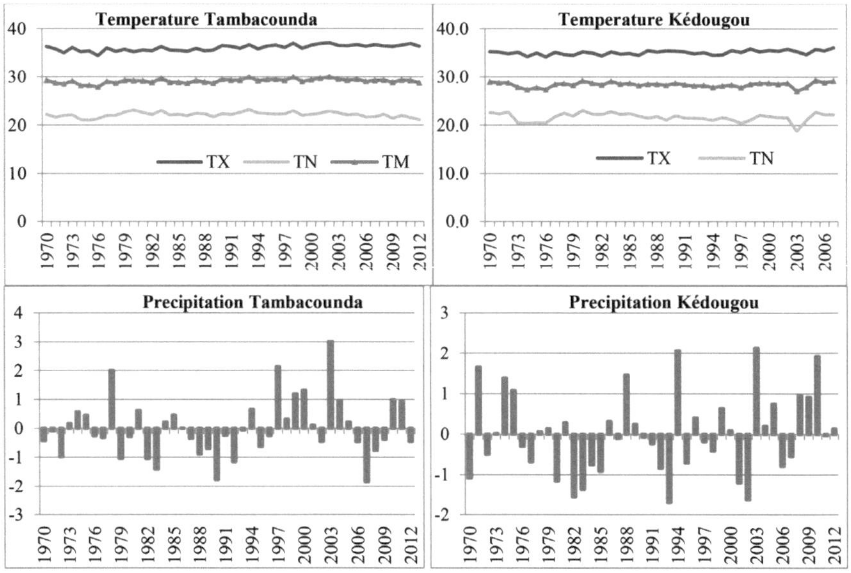

Figure 2.
Evolution of temperature and precipitation trend changes (1970–2012).

For precipitation, although both the Pettitt and Mann-Kendall tests showed no significant break-up and trend, with the p's for both tests being greater than the significance level of 0.01, the trend is slightly the rise from 1994 (**Figure 2**) and the rupture noted in 1994 and 2003 respectively in Tambacounda and Kedougou. Thus, the indices are generally negative from 1970 to 1994 and positive between 1994 and 2012, as illustrated by the positive Kendall τ with -0.074 mm at Tambacounda and 0.098 mm at Kédougou. This suggests that, beyond the drought of the 1970s, characterized by various works [7–10], another important change in rainfall patterns is still produced at the turn of the century as indicated by some works [13–17] who suggest the end of the Sahelian drought during the 1990s. Thus, on both sides of the year of rupture, rainfall increased by 13.3% in Tambacounda and 21% in Kédougou.

4.2 Impacts of climate change on surface waters in the basin

The impacts of climate change on surface water in the Gambia Basin have been analyzed using standardized flow indices. The application of the index method to the different stations allows a better comparison of the data of stations draining basins of sizes. These indices submitted to the Mann-Kendall and Pettitt tests revealed significant fluctuations with multiple consequences on the environment, hence the interest of studying them. The analysis of the tests shows the presence on the five selected stations of an upward trend and a break in 1994 (**Table 4**). However, the trend and the break were not significant at the 1% level. Thus, on both sides of 1994, two periods are established. The first one is from 1970 to 1994 and shows a trend of negative outflow relative to the drought of the 1970s [10] unlike the second (from 1994 to 2008) on which this trend, although not significant, is on the rise and corresponds to the improvement in rainfall conditions since the 1990s [14]. This upward trend in runoff in the Gambia Basin is in line with the work of Ali and Lebel [13] on the Sahelian zone, Ouoba [16] on Burkina Faso, Ozer et al. [14] on Niger, Niang [12] on Mauritania and Bodian [17] on Senegal.

Figure 3 shows the interannual variability of the standardized flow rates of stations in the Gambia Basin. It shows that this variability is almost synchronous

	Mann-Kendall test		Pettitt rest	
Gouloumbou	p-value	0.611	p-value	0.651
	τ of Kendall	−0.062	Date of rupture	1981
	Slope of Sen	−0.010	% Surplus or deficit	23.6
Mako	p-value	0.479	p-value	0.025
	τ of Kendall	0.094	Date of rupture	1994
	Slope of Sen	0.020	% Surplus or deficit	48.8
Simenti	p-value	0.241	p-value	0.011
	τ of Kendall	0.154	Date of rupture	1994
	Slope of Sen	0.030	% Surplus or deficit	64.7
Wassadou upstream	p-value	0.710	p-value	0.564
	τ of Kendall	−0.052	Date of rupture	1979
	Slope of Sen	−0.008	% Surplus or deficit	51.5
Wassadou downstream	p-value	0.065	p-value	0.002
	τ of Kendall	0.227	Date of rupture	1994
	Slope of Sen	0.026	% Surplus or deficit	44.6

(−): negative trend; (+): positive trend; (*): significant trend.

Table 4.
Results of Pettitt and Mann-Kendall runoff tests analyzed in the Gambia Basin (1953–2010).

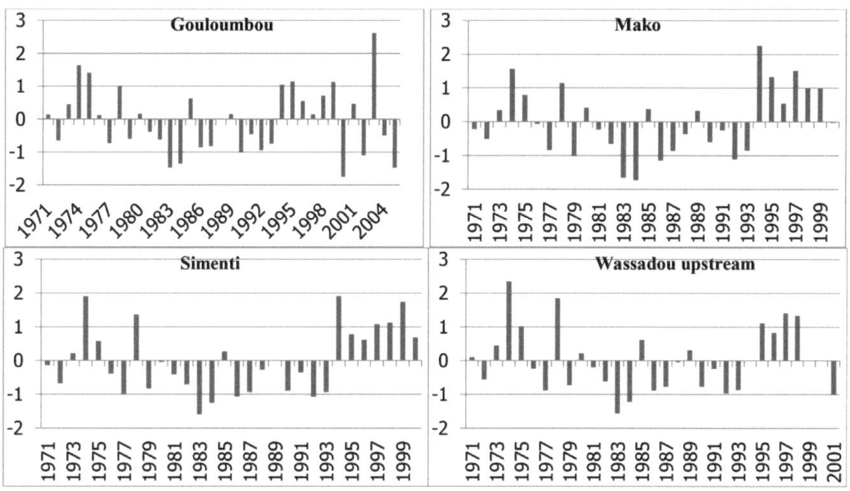

Figure 3.
Evolution of flow trend changes in the Gambia Basin (1970–2008).

with the presence of two hydroclimatic periods: a dry period between 1970 and 1994 and a dry period between 1994 and 2008. Thus, beyond the hydrological drought of the 1970s, a new hydrological change occurred again at the turn of the century (1990s), with river flows rising. On an annual scale, the evolution at the level of the stations is quite similar with the date of rupture which is noted in 1994. However, the surpluses are less important (23.6%) in Gouloumbou than on the other stations of the non where they exceed 50% (64.7% in Simenti, 51.5% in Wassadou Amont, 48.8% in Mako).

The causes of the tendencies and ruptures noted on the flow in the basin of the Gambia are sought on the various upheavals that this basin knew, the climate change in particular. One of its impacts was the severe drought of the 1970s [6, 7, 10, 14] which led to the decline in During the 1990s, the more or less favorable rainfall conditions were strictly responsible for this increase in runoff noted on the various stations of the study.

4.3 Impacts of climate change on groundwater in the basin

If the weakening of the low water levels during the 1970s resulted in a dwindling of the underground reserves of the river basins resulting from the accumulated rainfall deficits [23], the increase of the flow in the basin of the Gambia has at the same time caused an increase in the volumes drained by groundwater. The natural regime of low water on the Sudano-Sahelian rivers is very seriously affected by changes in climatic conditions. The drying up, by the volumes that it implies and its representativity of all the aquifers of the basin, constitutes an important characteristic of the hydrological regime of The Gambia. Drying data are given in **Tables 5** and **6**. Drying variables in the Gambia Basin at Mako and Simenti stations are highly variable.

The study of the drying coefficients shows that before 1994, year of rupture according to the Pettitt test (**Table 6**), a relative regularity of the values. Drying coefficients are generally low and range from 0.024 (in 2011–2012) to 0.076 (in 1977–1978) in Mako. For Simenti, they vary between 0.024 (in 2011–2012) and 0.074 (in 2004–2005). At the Mako station, the coefficients from 1970 to 1994, with an average value of 0.051, correspond to the lowest groundwater support volumes, the average volume of groundwater support being 3035 m^3/year over this period. On the other hand, over the period 1994–2014, the coefficients, with an average of 0.044, correspond to the support volumes of the highest water tables. Over this period, the average support volume for aquifers is 6531 m^3/year for a maximum of 16,454 m^3/year (in 2012–2013) in Mako. In the case of the Simenti station, the support volume of the aquifers ranges from 721 to 17,849 m^3, the average being 4138 m^3.

The analysis of **Tables 5** and **6** shows that, starting from the 1994 rupture, the rise in flows leads to a veritable drop in the values of the depletion coefficient, as evidenced by the negative character of Kendall's τ of the order of -0.11 in Mako and -0.06 in Simenti. This decrease in coefficients is responsible for an improvement in the volume of groundwater support in the general flow of the Gambia Basin, which resulted in a Kendall τ of the positive support volume (0.50 in Mako

	Station of Mako					**Station of Simenti**				
Descriptors	**Q_0**	**Q_t**	**t**	**α**	**V m^3/year**	**Q_0**	**Q_t**	**t**	**α**	**V m^3/year**
Mean	210	0.12	187	0.048	4612	203	0.30	171	0.05	4138
Standard deviation	109	0.33	15.0	0.01	2815	161	0.89	20	0.013	3427
Coefficient of variation	0.52	2.74	0.08	0.22	0.61	0.79	2.96	0.12	0.251	0.83
Minimum of the series	68	0.00	143	0.024	929	48.93	0.00	102	0.022	721
Maximum of the series	576	2.08	212	0.073	16,454	865	3.81	201	0.074	17,849

Q_0: flow at the beginning of the dry period; and Qt: flow at the end of the dry period; t: the number of days; α: the drying coefficient; V m^3/year: the support volume of the slicks.

Table 5.
Drying statistics in the Gambia Basin at Mako and Simenti (1970–2014).

	Mann-Kendall test						Pettitt test					
Mako	**p-value**	**τ of Kendall**	**Significance threshold**			**Sensitivity of trend**	**p-value**	**Date of rupture**	**Significance threshold**			**% Deficit or surplus**
			10%	**5%**	**1%**				**10%**	**5%**	**1%**	
Q0	<0.0001	0.44	Presence of trend			Rise	0.0001	1986	Presence of rupture			102
Qt	0.07	0.19	Presence of trend			Rise	0.0105	1994	Presence of rupture			285
t	0.06	0.20	Presence of trend			Rise	0.004	1985	Presence of rupture			9.1
K	0.29	−0.11	Presence of trend			Decline	0.10	1994	Presence of rupture			−15
V m^3/an	<0.0001	0.50	Presence of trend			Rise	<0.0001	1993	Presence of rupture			115
Simenti	**Mann-Kendall test**						**Pettitt test**					
	p-value	**τ of Kendall**	**Significance threshold**			**Sensitivity of trend**	**p-value**	**Date of rupture**	**Significance threshold**			**% Deficit or surplus**
			10%	**5%**	**1%**				**10%**	**5%**	**1%**	
Q0	0.44	0.09	Presence of trend			Rise	0.68	1994	Presence of rupture			−1.83
Qt	0.50	0.08	Presence of trend			Rise	0.11	1997	Presence of rupture			1887
t	0.25	−0.13	Presence of trend			Decline	0.23	1979	Presence of rupture			−4.28
K	0.59	−0.06	Presence of trend			Decline	0.13	1997	Presence of rupture			−21.3
V m^3/an	0.28	0.12	Presence of trend			Rise	0.23	1994	Presence of rupture			17.0

Table 6.
Results of the Pettitt and Mann-Kendall tests on drying variables in the Gambia Basin at Mako and Simenti stations (1970–2014).

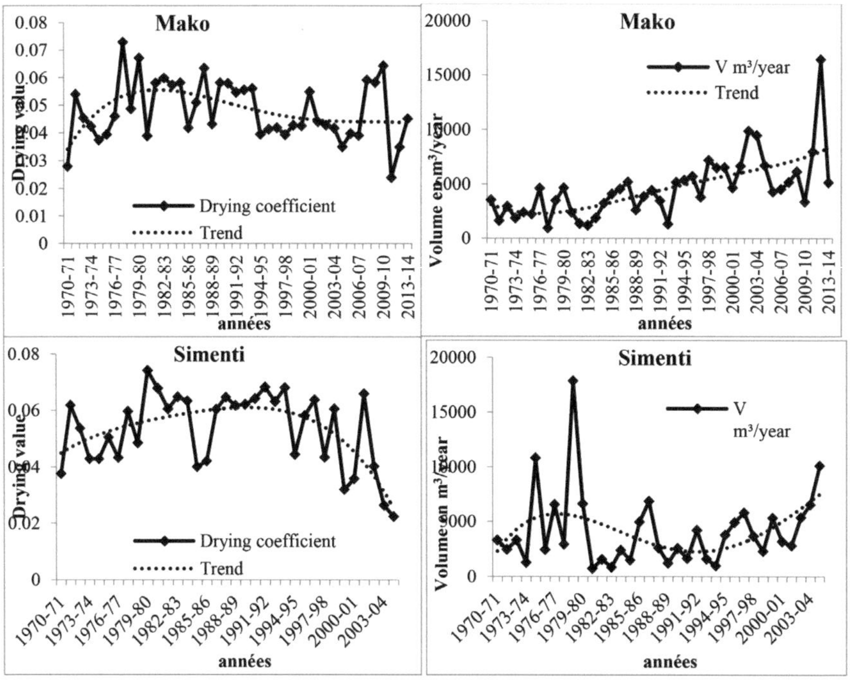

Figure 4.
Evolution of drying up in the Gambia Basin at the Mako and Simenti stations (1954–2014).

and 0.12 in Simenti). These results show that groundwater support volumes have increased in recent years in the basin. If this increase is only 16% in Simenti, in Mako, on the other hand, it is 115%.

This extremely important variation in the increase in water volumes mobilized by aquifers (115% in Mako and only 17% in Simenti) is explained, and in a normal way, by the difference in the study period. In fact, given available data, this study period is longer in Mako (1970–1971 and 2013–2014) than in Simenti (1970–1971 and 2004–2005), while the most recent years (2005–2006 to 2013–2014) are largely rainy and most likely to feed the water table of basin. Thus, on both sides of rupture that occurred in 1993 in Mako, the average volume of groundwater support has more than doubled, from 3035 m^3 over period before rupture (1970–1971 and 1993–1994) at 6531 m^3 over period after rupture (1994–1995 and 2013–2014). On the other hand, in Simenti, where the rupture occurred in 1994, it only went from 3916 m^3 in period before rupture (1970–1971 and 1994–1995) to 4581 m^3 in period before rupture (1995–1996 and 2004–2005).

Figure 4 compares the evolution of the depletion coefficient and the support volume of the aquifers in the basin. Both variables move in opposite directions. The decrease in the dry-season coefficient in the current dry period essentially corresponds to an increase in the extent and width of the water table in the basin. The period 1994–2014, compared to that of 1970–1993, is in surplus for the duration of the drying up and for the volumes of support, and deficit for the coefficient of dryness.

5. Discussion

The characterization of climate change and its impacts on water resources in the Gambia watershed over the period 1970–2014, based on indices of rainfall, flow, and drying coefficients, highlighted two more or less opposite periods: the first period generally going from 1970 to 1994 and the second period generally between 1994 and 2014.

In the climate domain, the analysis of the series of minimum, maximum and average temperatures used indicated a warming trend from the 1990s. This rise in temperatures (around 2.61% for TX and 1.35% for TN at Tambacounda) is in concert with the results of numerous studies [30–32] that have highlighted a context of global warming. For the rainfall analysis, the standardized rainfall indices calculated at the Tambacounda and Kedougou stations are generally negative from 1970 to 1994 and positive between 1994 and 2012. Although the rainfall decline has intensified in the Gambia Basin during in the 1970s and 1980s, it did not persist in the 1990s and 2000s, as noted by some authors [33–35]. Thus, between the periods 1970–1994 and 1994–2012, rainfall in the basin increased by 13.3% in Tambacounda and 21% in Kédougou. These results confirm the dryness of the 1970s and 1980s in many studies [4–10] and the significant new change in rainfall patterns that occurred in the 1990s as indicated by some works [13–17].

To characterize the impacts of climate change on surface water resources in the Gambia Basin, standardized flow indices are analyzed. The results show a break in 1994 and an upward trend in the flow in the basin, unlike in the 1970s and 1980s, where discharges in the basin declined significantly. This resulted in no less significant surpluses (23.6% in Gouloumbou 50%, 64.7% in Simenti, 51.5% in Wassadou Amont, 48.8% in Mako). This upward trend noted in the Gambia Basin is consistent with the work of Ali and Lebel [13] on the Sahelian zone, Ouoba [16] on Burkina Faso, Ozer et al. [14] on Niger, Niang [12] on Mauritania and Bodian [17] on Senegal. Numerous studies carried out in West and Central Africa have highlighted this decrease in surface and groundwater flow as a result of the drop in rainfall [4–6] and this increase in runoff from the 1990s [14–16], foreshadowing an improvement in the hydrological regime in this area.

These hydrological effects of climate change on surface and groundwater in Gambia River Basin can be extrapolated to other basins. Thus, a relationship is established between increase in discharges and volumes of groundwater support in the Gambia River Basin over recent period with that noted in other basins on national territory, such as Senegal River Basin [10] of the subregion, such as the Sahelian basins [36, 37] and the world, such as the basins of the South American rivers [38], China [39], Finland [40], from 48 contiguous US states [41], on which flow has increased.

To characterize the impacts of climate change on the groundwater resources of the Gambia Basin, the drying coefficients at the Mako and Simenti stations are analyzed. The results, as well as the increase in rainfall and run-off in the basin, a downward trend of the drying-out coefficients (−15% in Mako, −23% in Simenti) and the increase in support volumes of groundwater (115% in Mako, 17% in Simenti) is noted in the Gambia Basin from 1994. These results confirm the increase in water volumes mobilized by aquifers linked to the increase in rainfall observed over the years 1990. Although during the period 1970–1994 there was a considerable reduction in the underground reserves that normally supply dry-season streams [42–45], from 1994, the improvement of the rainfall conditions led to an increase in the volume of water mobilized by the aquifers, suggesting a progression of the underground reserves and an increase of the flows.

6. Conclusion

The purpose of this article was to evaluate surface water resources by calculating normalized indices of rainfall and flow, and groundwater from the Maillet model in a climate change context. The results show that the rain, the water slide and the volumes mobilized by the aquifers in the Gambia Basin increase from 1994. For the rain, this increase is 16.1% Kédougou and 13.3% to Tambacounda. For flows flowing into the Gambia Basin, the same increase is costatée, an increase that can sometimes exceed 50% at certain stations (48.7% in Gouloumbou, 64.6% in Mako, 34.7% in Simenti 85.3 at Wassadou Amont). As for the drying coefficients, they vary between 0.051 and 0.044 at the Mako station (a decrease of −15.1%) and between 0.057 and 0.045 at the Simenti station (a decrease of −21.3%). This resulted in longer drying times and increased water volumes mobilized by aquifers in the order of 115% in Mako and 17% in Simenti. These results highlight an increase in volumes of water drained and mobilized by aquifers after 1994, and suggest an increase in surface and groundwater resources under the influence of climate change.

This increase in rainfall and recorded flows in the basin augurs the improvement of rainfall patterns in the area compared to the drought period of previous decades, although the persistence and sustainability of the current increase is still to be proven, knowing that the climatological scale, par excellence is the thirties. The study of the hydrological variability of the Gambia River Basin is therefore perceived as a major element to be taken into account for a better understanding of the hydrological response of major watersheds to climate change. To better determine the impacts of climate change on water resources in the basin, trend analysis needs to be combined with other approaches, such as surface condition analysis. It would therefore be important to know the dynamics of land use, in a context of climate change, and its impact on the flow in the basin, which is now possible with remote sensing techniques based on satellite imagery.

Author details

Cheikh Faye
Department of Geography, U.F.R. Science and Technology, UASZ, Laboratory of Geomatics and Environment, Ziguinchor, Sénégal

Address all correspondence to: cheikh.faye@univ-zig.sn

IntechOpen

© 2019 The Author(s). Licensee IntechOpen. This chapter is distributed under the terms of the Creative Commons Attribution License (http://creativecommons.org/licenses/by/3.0), which permits unrestricted use, distribution, and reproduction in any medium, provided the original work is properly cited.

References

[1] IPCC. Résumé à l'intention des décideurs. In: Solomon S, Quin A, Manning M, Chen Z, Marquis M, Averyt KB, Tignor M, Miller HL, editors. Changements climatiques 2007: Les éléments scientifiques, Contribution du Groupe de travail I au quatrième Rapport d'évaluation du Groupe d'experts intergouvernemental sur l'évolution du climat; Cambridge, UK/ New York, NY, USA: Cambridge University Press; 2007. 18p

[2] Faye C, Diop EHS, Mbaye I. Impacts des changements de climat et des aménagements sur les ressources en eau du fleuve Sénégal: Caractérisation et évolution des régimes hydrologiques de sous-bassins versants naturels et aménagés. Belgeo. 2015;**4**:1-22

[3] Kanohin F, Saley MB, Savané I. Impacts de la variabilité climatique sur les ressources en eau et les activités humaines en zones tropicale humide: Cas de la région de Daoukro en Côte d'Ivoire. European Journal of Scientific Research. 2009;**26**(2):209-222

[4] Dione O. Evolution climatique récente et dynamique fluviale dans les hauts bassins des fleuves Sénégal et Gambie [Thèse de doctorat]. Université Lyon 3 Jean Moulin; 1996. 477p

[5] Sighomnou D. Analyse et redéfinition des régimes climatiques et Hydrologiques du Cameroun: Perspectives d'évolution des ressources en eau [Thèse Doctorat d'Etat]. Université de Yaoundé 1, Département des Sciences de la Terre; 2004. 291p

[6] Goula BTA, Savane I, Konan B, Fadika V, Kouadio GB. Comparative study of climatic variability impact on water resources of N'zo and N'zi watersheds in Côte d'Ivoire. Sciences & Nature. 2005;**2**(1):10-19

[7] Sow AA. L'hydrologie du Sud-est du Sénégal et de ses Confins guinéo-maliens: Les bassins de la Gambie et de la Falémé [Thèse (PhD)]. Université Cheikh Anta Diop de Dakar; 2007. 1232p

[8] Bodian A. Approche par modélisation pluie-débit de la connaissance régionale de la ressource en eau: Application au haut bassin du fleuve Sénégal [Thèse de doctorat]. UCAD; 2011. 288p

[9] Faye C. Evaluation et gestion intégrée des ressources en eau dans un contexte de variabilité hydroclimatique: Cas du bassin versant de la Falémé [Thèse (PhD)]. Université Cheikh Anta Diop de Dakar; 2013. 309p

[10] Faye C, Sow AA, Ndong JB. Étude des sècheresses pluviométriques et hydrologiques en Afrique tropicale: Caractérisation et cartographie de la sècheresse par indices dans le haut bassin du fleuve Sénégal. Physio-Géo - Géographie Physique et Environnement. 2015;**9**:17-35

[11] Ali A, Lebel T, Amami A. Signification et usage de l'indice pluviométrique au Sahel. Sécheresse. 2008;**19**(4):227-235

[12] Niang AJ. Les processus morphodynamiques, indicateurs de l'état de la désertification dans le Sud-Ouest de la Mauritanie. Approche par analyse multisource [Thèse de Doctorat]. Belgique: Université de Liège; 2008. 286p

[13] Ali A, et Lebel T. The Sahelian standardized rainfall index revisited. International Journal of Climatology. 2009;**29**(12):1705-1714

[14] Ozer P, Hountondji YC, Laminou MO. Évolution des caractéristiques

pluviométriques dans l'Est du Niger de 1940 à 2007. Revue Internationale de Géologie, de Géographie et d'Ecologie Tropicales. 2009;**33**:11-30

[15] Hountondji YC, Sokpon N, Nicolas J, Ozer P. Ongoing desertification processes in the sahelian belt of West Africa: An evidence from the rain-use efficiency. In: Röder A, Hill J, editors. Recent Advances in Remote Sensing and Geoinformation Processing for Land Degradation Assessment. CRC Press, série ISPRS. 2009:**8**:173-186

[16] Ouoba AP. Changements climatiques, dynamique de la végétation et perception paysanne dans le Sahel burkinabè [Thèse de Doctorat Unique]. Burkina Faso: Université de Ouagadougou; 2013. 305p

[17] Bodian A. Caractérisation de la variabilité temporelle récente des précipitations annuelles au Sénégal (Afrique de l'Ouest). Physio-Géo [En ligne]. 2014;**8**(1):297-312. DOI: 10.4000/physio-geo.4243. URL: http://physio-geo.revues.org/4243

[18] Lamagat JP. Monographie Hydrologique du Fleuve Gambie Collection M&m, ORSTOM-OMVG; 1989. 250p

[19] Mckee TB, Doesken NJ, Kleist J. The relationship of drought frequency and duration to time scale. In: Actes de la 8th Conference on Applied Climatology; Anaheim, Californie; 1993. pp. 179-184

[20] Hayes M. Drought Indices. National Drought Mitigation Center. 1996. Disponible sur: http://enso.unl.edu/ndmc

[21] Jouilil I, Bitar K, Salama H, Amraou I, Mokssit A, Tahiri M. Sécheresse météorologique au bassin hydraulique OUM ER RBIA durant les dernières décennies. Larhyss Journal. 2013;**12**: 109-127

[22] Sharma TC, Panu US. Analytical procedures for weekly hydrological droughts: A case of Canadian rivers. Hydrological Sciences Journal. 2010; **55**(1):79-92

[23] Briquet JP, Mahé G, Bamba F. Changements climatiques et modification du régime hydrologique du fleuve Niger à Koulikoro (Mali). In: Actes de Conférence de Paris. Paris: IAHS Publication; 1995. pp. 113-124

[24] Faye C. Caractérisation des basses eaux: Les effets durables du déficit pluviométrique sur les étiages et le tarissement dans le bassin du bakoye. In: Revue Espaces et Sociétés en Mutation. Presses Universitaires de Dakar – Numéro Spécial – 2015. 2015. pp. 109-126

[25] Pettitt AN. A non-parametric approach to the change-point problem. Applied Statistics. 1979;**28**(2):126-135

[26] Mann HB. Nonparametric tests against trend. Econometrica. 1945;**13**(3): 245-259

[27] Kendall M. Multivariate Analysis. London: Charles Griffin & Company; 1975. 202p

[28] Hirsch RM, Slack JR. A nonparametric trend test for seasonal data with serial dependence. Water Resources Research. 1984;**20**:727-732

[29] Yue S, Wang CY. The Mann-Kendall test modified by effective sample size to detect trend in serially correlated hydrological series. Water Resources Management Journal. 2004; **18**:201-218

[30] Leroux M. Dynamic Analysis of Weather and Climate, Atmospheric Circulation, Perturbations, Climate Evolution. UK: Springer/Praxis Publishing; 2010. 422p

[31] Sagna P. Important refroidissement du temps à Dakar dans un contexte de réchauffement climatique : analyse de la situation météorologique des 16 et 17 février 2014. Annales de la Faculté des Lettres et Sciences Humaines. 2013;**43** (B):95-117

[32] IPCC. Changements climatiques 2013. Les éléments scientifiques. In: Résumé à l'intention des décideurs, OMM, PNUE; 2013. 14p

[33] L'Hôte Y, Mahé G, Some B, Triboulet JP. Analysis of a sahelian rainfall index from 1896 to 2000: The drought continues. Hydrological Sciences. 2002;**47**(4):563-572

[34] Frappart F, Hiernaux P, Guichard F, Mougin E, Kergoat L, Arjounin M, et al. Rainfall regime across the Sahel band in the Gourma region, Mali. Journal of Hydrology. 2009;**375**:128-142

[35] Soro GE, Anouman DGL, Goula BITA, Srohorou B, Savane I. Caractérisation des séquences de sécheresse météorologique a diverses échelles de temps en climat de type soudanais: Cas de l'extrême Nord-ouest de la cote d'ivoire. Larhyss/Journal. 2014;**18**:107-124

[36] Amogu O, Descroix I, Souley Yero K, Le-breton E, Mamadou I, Ali A, et al. 2010. Increasing river flows in the Sahel?. Water 2 (2) 170–199

[37] Mahe G, Paturel J-E. 1896–2006 Sahelian annual rainfall variability and runoff increase of Sahelian rivers. Comptes Rendus, Geoscince. 2009; **341**(7):538-546

[38] Callede J, Guyot JL, Ronchail J, L'Hote Y, Niel H, de Oliveira E. Evolution du débit de l'Amazone à Óbidos de 1903 à 1999, Evolution of the River Amazon's discharge at Óbidos from 1903 to 1999. Hydrological Sciences Journal. 2004;**49**(1):85-97

[39] Tao WK, Simpson J. The Goddard cumulus ensemble model. Part I: Model description. Terrestrial, Atmospheric and Oceanic Sciences. 1993;**4**:35-72

[40] Hyvärinen V. Observed trends and fluctuations in hydrological time series in Finland—A review. In: Proceedings of the Second International Conference on Climate and Water; Espoo, Finland; 1998. pp. 1064-1070

[41] Walter MT, Wilks DS, Parlange JY, et al. Increasing evapotranspiration from the conterminous. United States Journal of Hydrometeorology. 2004;**5**: 405-408

[42] Vissin WE. Impact de la variabilité climatique et de la dynamique des états de surface sur les écoulements du bassin béninois du fleuve Niger [Thèse de Doctorat]. France: Université de Bourgogne; 2007

[43] Mahé, G, Dessouassi R, Bandia C and Olivry JC. Comparaison des Fluctuations Interannuelles de Piézométrie, Précipitation et débit sur le Bassin Versant du Bani à Douna au Mali. IAHS Publication; 1998

[44] Olivry JC, Bricquet JP, Mahé G. Variabilité de la Puissance des crues des Grands Cours d'eau d'Afrique Intertropicale et Incidence de la Baisse des Écoulements de base au Cours des deux Dernières Décennies. IAHS Publication; 1998

[45] Amoussou E. Variabilité pluviométrique et dynamique hydro-sédimentaire du bassin-versant du complexe fluvio-lagunaire mono-ahémé-couffo (Afrique de l'Ouest) [Thèse de Doctorat]. Université de Bourgogne, Centre de Recherche de Climatologie Bahire S. 1986. Monographie sur le N'zi, direction de l'hydraulique Humaine; 2010

Chapter 5

Analysis of Non-Rainfall Periods and Their Impacts on the Soil Water Regime

Milan Gomboš, Branislav Kandra, Andrej Tall and Dana Pavelková

Abstract

Rainfall and evaporation belong to the basic components of the hydrological cycle. Rainfalls are a decisive natural source of water in the soil. For water replenishment in the natural environment, it is important not only the sum of the rainfall for the balanced period but also the time distribution. In the case of long non-rainfall periods, the soil profile is dried. In the sufficiently long non-rainfall period, the water reserves in the unsaturated zone of the soil profile change and the actual evapotranspiration is reduced. There is a meteorological and then dry soil formation. For the design of adaptation measures, it is necessary to quantify the mentioned hydrological processes. These were investigated in the central area of the Eastern Slovak Lowland between 1970 and 2015. Significant non-rainfall periods, their periodicity and statistical characteristics have been identified. In the course of significant non-rainfall intervals during the vegetation periods the water reserves in the root layer of the soil were analysed up to a depth of 1 m, the actual and potential evapotranspiration, the evapotranspiration deficit, the groundwater level and the air temperature. The longest non-rainfall periods exceeded 30 days.

Keywords: rainfall, evaporation, non-rainfall periods, evapotranspiration deficit, soil drought

1. Introduction

Precipitation and evapotranspiration are basic compounds of hydrologic cycle. Precipitation is crucial natural source of water in the soil. Precipitation amount and temporal distribution of the rainfall is important for water refilling of the environment for balanced periods. Drying of soil profile occurs during long rainless periods. Meteorological drought and subsequently soil drought occurs in the case of the sufficiently long rainless period. Therefore it is necessary to know size and statistical characteristics of rainless periods (RLP).

For the analysis of the impacts of the length of the non-precipitation period on the soil water regime, it is necessary to analyse the second critical component of the hydrological cycle. It is a process of evaporation and quantification of the vapour. The evaporation of water in nature is physically the process of converting water to vapour. Due to the high vapour heat (2450 kJ kg^{-1} at the evaporating surface temperature of 20°C) it is an energy-intensive process. In nature, it is the decisive regulator of energy flows. The evaporation from the soil and plants is called evapotranspiration. The intensity of evapotranspiration over time is affected by the power consumption required for the phase conversion of water to vapour and the amount of this water in the environment. Consequently, evapotranspiration is seasonal in the course of the year. The decisive amount of water evaporates in the vegetal half of the year—from April to September. The maximum possible evapotranspiration under given meteorological conditions is potential evapotranspiration (ET_0). Its size is the result of the energy balance on the active surface. Really evaporated water from the soil and plants is called the actual evapotranspiration (ET_a). It depends on the positive energy balance, the turbulent exchange between the plant and the atmosphere, the water content of the soil profile, and the plant's ability to regulate the intake and discharge of water from its own organism. If there is sufficient water in the environment, then the actual evapotranspiration is the same as the potential evapotranspiration. If $ET_0 > ET_a$, there is an evapotranspiration deficit. This in the root zone of the soil profile indicates the water supply deficit and the beginning of the soil profile drying. Current climate changes are reflected in the redistribution of precipitation over the course of the year and the increase in the air temperature. In other words, the distribution of water and energy in the hydrological system is redistributed. These changes are reflected in changes in evaporation and consequently in changes in the water balance of the area [1–6]. The authors of the chapter have the hypothesis that not only the height of precipitation but also their time distribution during the year has the effect on the occurrence of soil drought.

The aim of the chapter is to identify significant non-precipitation periods, to quantify their time duration, probability properties, development trends and impacts on changes in the critical components of the water regime in the soil environment. In the course of significant non-precipitation intervals during vegetation periods, the water reserves in the root layer of the soil up to a depth of 1 m, the actual and potential evapotranspiration, the evapotranspiration deficit, the groundwater level and the air temperature were analysed.

2. Materials and methods

The identification of non-precipitation periods and the process of their impact on the components of the water regime of the soil environment were examined in the Milhostov locality in the Eastern Slovak Lowlands (ESL).

The analysis concerned the growing seasons from 1970 to 2015. During the analysis, the following values were analysed at 1-day calculation step and one-day input data: non-rainfall periods, soil water storage to the depth of 1 m; movement of the groundwater level; air temperature; and daily totals of actual and potential evapotranspiration, rainfall and evapotranspiration deficiency.

The research was based on field measurements of hydrophysical characteristics, volume humidity monitoring over the soil profile vertical line, laboratory work and numerical simulation of the water regime using GLOBAL mathematical model [1–6]. The part of the research works was also the verification of the mathematical model using results of the monitoring carried out in the examined locality.

2.1 Description of the investigated area

These hydrological processes were examined in Milhostov, the area located in the central part of the ESL, Slovakia (48°40′11.08″, N; 21°44′18.02″, E; 100 m). The selected area is characteristic of the ESL. Typical soil of the area is gleyic fluvisol, a medium-heavy soil where clay particles forms 11–39% of the content.

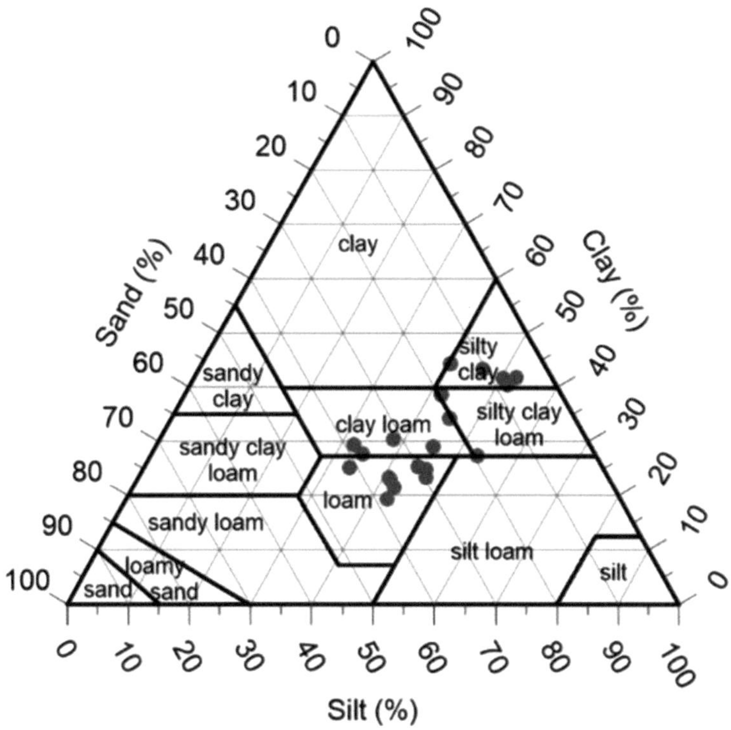

Figure 1.
Texture of the examined soil profile as defined by the USDA triangle (sand 0.05–2.0 mm, clay <0.002 mm, silt 0.002–0.05 mm).

The profile is made of clay, clay-loam and silty clay layers **Figure 1**.

Figure 2 shows the course of the clay, dust and sand content via the vertical line of the soil profile by 0.1 m layers to a depth of 1.0 m.

The clay, silt and sand content via the vertical line of the soil profile to 1 m depth. **Figure 2** shows that the most ratio of the studied profile is formed by the dusty component. The profile is up to 0.7 m relatively homogeneous. In the range between 0.7 and 1.0 m, the ratio of the individual components of the microstructure significantly changes. For this reason, two material layers were considered in the calculations. **Table 1** lists their retention lines parameters according to Van Genuchten.

In terms of climate, the examined area, as well as the rest of the ESL, is located in the transitional climate region between the maritime and the continental climate. In terms of temperature, the area is homogeneous. Long-term mean temperature in the area between the years 1961 and 2015 is 9.4°C; minimum mean daily temperature is −20.5°C, maximum mean daily temperature is +30.6°C. The absolute minimum temperature in the area during the analysed period was −29.1°C (January 01, 1987) and the absolute maximum daily temperature was +38.2°C (July 22, 2007). The warmest month is July, the coldest is January. The mean annual amount of rainfall in the area is 558 mm (years 1961–2015). Daily maximum rainfall total was 82.5 mm (June 26, 1995). Rainfall conditions in the area are heavily influenced by the air circulation. Heaviest rainfall is induced by the humid and warm air flowing from the south. Air currents from the other directions do not usually bring rainfall because it falls on the Carpathians. The maximum duration of the continuous period with no rainfall was 35 days. The maximum annual long-term mean duration of the period with no rainfall is 17 days. Long-term mean

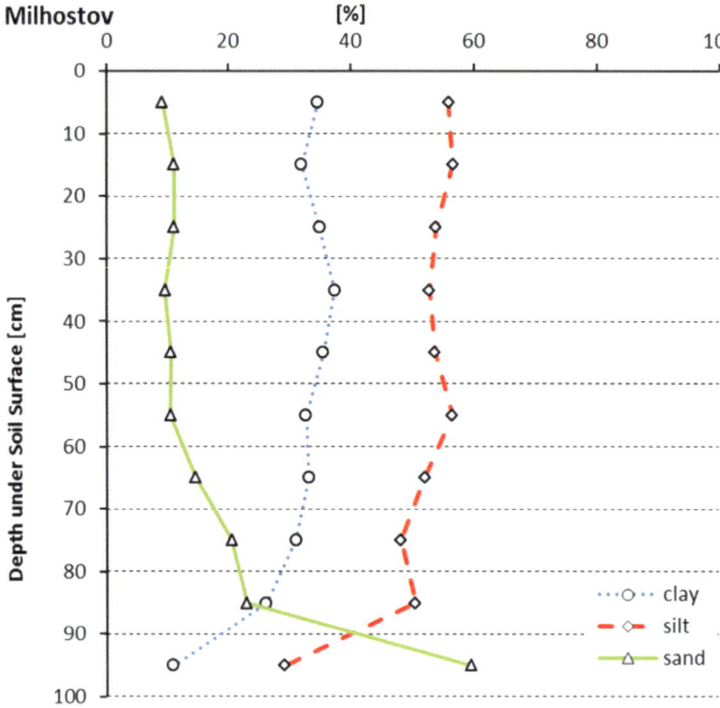

Figure 2.
Content of clay, silt and sand to the soil profile vertical to 1 m depth.

Layer	Alpha	n	θs	Θr
0.0–0.7 m	0.0084	1.5202	0.43	0.12
0.7–1.0 m	0.0482	1.1593	0.39	0.07

Table 1.
Parameters of retention lines of material layers of the soil profile.

wind speed 10 m above the ground is 2.468 m s^{-1}. Daily long-term mean pressure of water vapour is 0.999 kPa where the daily mean minimum is 0.083 kPa and the maximum is 2.760 kPa. Daily long-term mean relative humidity is 77.4%, where the daily mean minimum is 33% and the maximum is 100%. With regard to phenological characteristics and vegetation of the area, there are no noticeable temporal or local changes.

2.2 Description of the experiment

Methodical process of the research consists of two phases: rainless periods selection, model verification, calculation of water regime components and results interpretation. Climatic station of Milhostov (N 48°39,786′; E 21°43,298′) was chosen for selection of statistically important rainless periods. Station is located at central part of Eastern Slovak Lowland (ESL) and represents wider area of lowland. Daily precipitation amounts of the 1961–2015 period was examined for the station. 20,080 daily precipitation amounts (including zero) were analysed for the period. Length of RLP was identified separately for whole years of examined periods and separately for vegetal periods (VP). Selections of RLP were applied for VP examination in two ways. Periods with zero daily precipitation amounts were considered in first selection (s0). Daily precipitation amounts lower than 2 mm were considered as zero in the second selection (s2). These precipitation are likely to be catched by the plant cover and subsequently vaporised without influencing soil water supply. Selection of rainless periods during VP was done only of those which occurred during vegetal periods (April–September). However, periods that starts before VP and overlap to VP and those that starts during VP and finished after VP were considered as well. Rainless periods of the 1961–2015 were probability-evaluated in every selection group.

Just (s0) selection was applied during examination of entire year periods. Basic statistical characteristics were calculated after identification of RLP. Probability characteristics of occurrence of different RLP temporal duration were calculated by binomial distribution. The results are indicated in the form of frequency curves. Development of number of rainless days during individual years and VP was evaluated. For example, 10 longest RLP during 1961–2015 were selected for every selection.

The database for needs of the experiment was collected by the field monitoring, laboratory measurements and the numerical simulation via the mathematical model 'GLOBAL'.

The field monitoring concerned the measuring of the groundwater level and volumetric moisture within a soil profile, vertically to the depth of 0.8 m and horizontally by 0.1 m thick layers. The field works included also soil sampling. Gathered soil samples were processed in the laboratory and basic characteristics of the soil profile were determined. The Climatic and Agroecology Research Institute is located in the examined area. During the years 1970–2015, the institute provided hydro-meteorological data and plant characteristics necessary for the numerical simulation via the mathematical model 'GLOBAL'. The results of the water storage monitoring to the depth of 0.8 m are available as well. The model was verified using the data from the extremely dry growing season in 2007. The growing season of 2007 was the second driest season of the analysed period. The driest season was that of the year 2015.

Once the model was verified, the development of ET_0, ET_a and water storage in a soil profile to the depth of 1 m was calculated with one-day calculation step. In addition, the value of evapotranspiration deficiency was calculated using Eq. (1):

$$D = ET_0 - ET_a \tag{1}$$

The simulation was conducted for the growing seasons from 1970 to 2015. ET_0, ET_a and D totals during the growing seasons were analysed with regard to water storage (WS), rainfall and groundwater level (GWL). For ET_0, ET_a and D, linear trends were calculated and the correlation analysis between the analysed units was performed. The development of *WS* in the extremely dry season was analysed with one-day step.

2.3 Brief description of the model 'GLOBAL'

The model 'GLOBAL' is a mathematical model to enable simulating water movement in soil and calculating the distribution of soil moisture potential, i.e. soil moisture in real time [7–12]. The model is based on the numerical solution of a non-linear partial differential Richards' equation describing water movement in aerated soil zone, which is as follows:

$$\frac{\partial h_w}{\partial t} = \frac{1}{c(h_w)} \frac{\partial}{\partial z} \left[k(h) \left(\frac{\partial h_w}{\partial z} + 1 \right) \right] - \frac{S(z,t)}{c(h_w)} \tag{2}$$

where h_w—soil moisture potential; z—vertical coordinate; $k(h_w)$—unsaturated hydraulic conductivity of soil; $S(z,t)$—intensity of water uptake by plant roots from unit soil volume per unit of time (cm^3/cm^3).d^{-1}; and $c(h_w) = \frac{\partial \theta}{\partial h_w}$, θ—volume soil moisture (cm^3/cm^3).

The simulation using the model 'GLOBAL' can be performed with one-day calculation step. In that instance, daily values are used as the input values for creating the boundary conditions. This also applies to the meteorological inputs and the plant cover input parameters. The model 'GLOBAL' includes also soil hydrophysical characteristics such as: retention curves, soil saturated and unsaturated hydraulic conductivity, hydrolimits and some physical soil characteristics (porosity, specific weight and volumetric mass density and moisture of saturated soil). In the model, hydrophysical characteristics are expressed by analytical equations. Moisture retention curve is described by van Genuchten model.

Potential evapotranspiration ET_0 is calculated by Penman-Monteith equation. For determining actual transpiration and evaporation, the method developed at IH SAS was used. This method assumes that the evapotranspiration depends on the value of leaf area index (LAI). The intensity of potential evaporation E_{eo} is calculated from the value of potential evapotranspiration ET_0 as follows:

$$E_{eo} = ET_0 . \exp(-m_1 . LAI) \tag{3}$$

The value of empirical coefficient ($m_1 = 0.463$) was gained by field measurements in a maize field. Actual evapotranspiration and its structure is calculated based on the values of potential evapotranspiration ET_0 and the relation between relative evapotranspiration E_{eo}/ET_0 and soil profile moisture, i.e.

$$ET_r = \frac{E_{eo}}{ET_0} = f(\theta) \tag{4}$$

The calculation method is based, inter alia, on the assumption that the median soil moisture in the root zone used for the calculation of transpiration, or the value of moisture in the upper layer of a soil profile used for the calculation of evaporation, depends on the intensity of evaporation. The higher is the intensity of evaporation, the higher is the value of θ_k, at which the evaporation starts to decrease. The verification of the method via the model 'GLOBAL' has shown a very high concordance between the calculated values and the values measured in real conditions. The outputs of the modelling are moisture and soil moisture potential distribution, daily totals of the following: interception, evaporation and its components, infiltration, existing water deficiency in soil and other.

3. Results and discussion

3.1 Identification of non-precipitation periods

Basic characteristics of descriptive statistics are indicated in **Table 2**. It results that 12,185 days (60.68%) were with no precipitation in 1 year selection data series. Days without rainfall were grouped to 3409 periods. 63.87% of days without rainfall were in the case of VP and s0 selection and 85.12% days without rainfall in the case of VP and s2 selection. The longest RLP duration was 35 days for s0 selection and 53 days for s2 selection.

Eloquent image of probability of occurrence of maximal year durations of RLP gives probability of exceedance curves. Curves are indicated in **Figure 3**. There of

Parameters	Year	VP	VP
Selection	f(0)	f(0)	f(2)
Mean	3.57	3.67	6.15
Standard error	0.06	0.08	0.17
Median	2	2	4
Mode	1	1	1
Standard deviation	3.52	3.51	6.31
Sample variance	12.41	12.30	39.84
Kurtosis	10.98	9.66	8.47
Skewness	2.64	2.47	2.45
Range	34	34	52
Minimum	1	1	1
Maximum	35	35	53
Sum	12,185	6429	8567
Count	3409	1751	1393
Confidence level (95.0%)	0.12	0.16	0.33

Table 2.
Basic statistical characteristics of rainless periods.

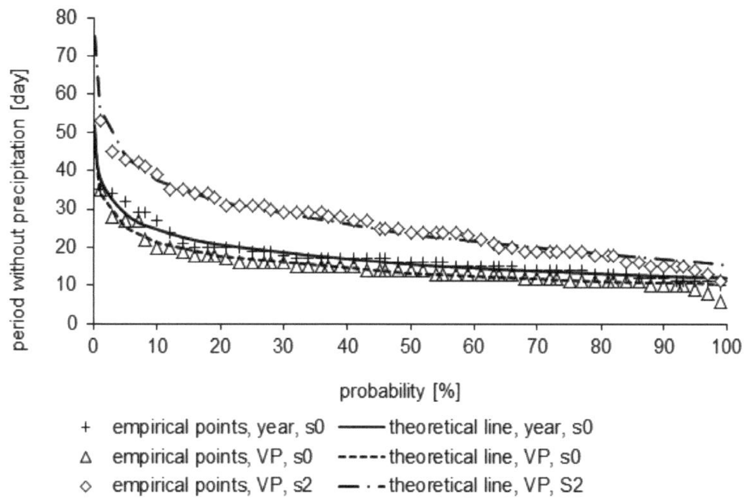

Figure 3.
Empirical and theoretical probability of exceedance curves of maximal year durations of rainless periods.

results that only little differences are between course of yearly and vegetal selections of s0. Length of RLP with p = 50% probability of exceedance is for this selection of VP 13 days and for the year 15.5 days. This value is 23.3 days for selection s2 during the vegetal period. The course of probability of exceedance curve is in case of s2 selection moved and more steep. This indicates greater variability and longer duration of RLP.

Numbers of days without rainfall are indicated in **Figure 4** for individual years and their vegetal periods for selections s0 and s2. Developmental trends are for all cases parallel with timeline. It is testified by the fact that number of rainy

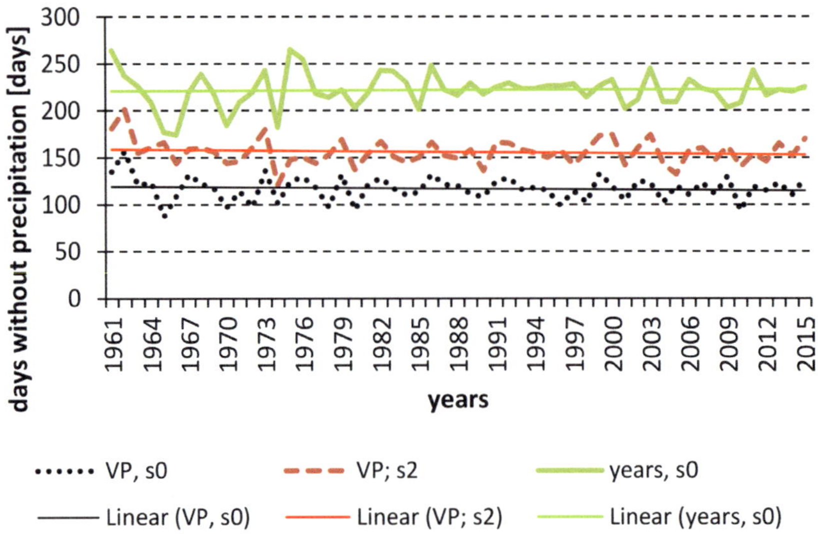

Figure 4.
Number of days without precipitation in individual years 1961–2015.

and rainless days is for individual years and their vegetal periods balanced in the long term view. Only duration of rainless events of non-vegetal periods changes in examined periods. It is confirmed by the results that are graphically shown in **Figure 5**. There are indicated numbers of days of rainless periods that lasts 10 days or more for VP of individual years. From the figure results that trend development of both selections s0 and s2 is stable without change (parallel to the timeline) during vegetal periods. Elongated trend of rainless periods with duration of 9 days or more was identified in case of s0 yearly selection. These results show that this trend

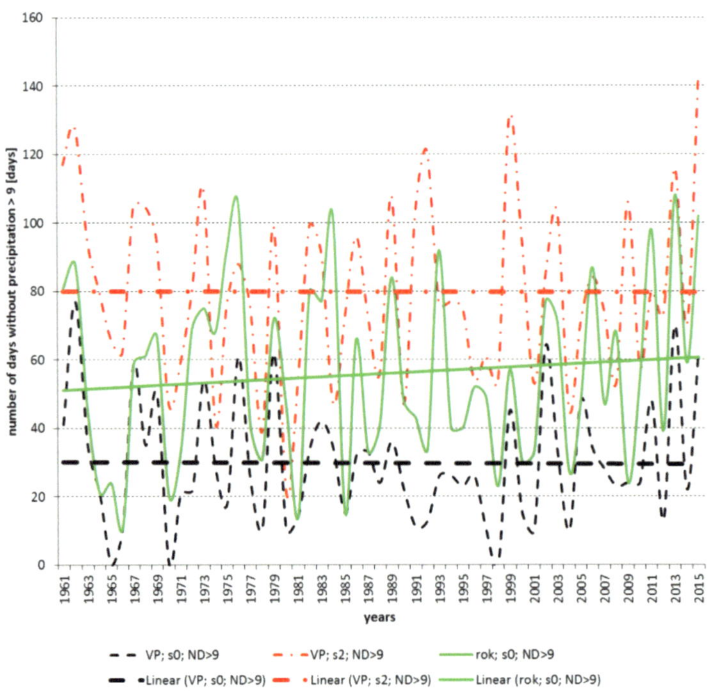

Figure 5.
Number of days during rainless periods with duration of 9 days or more.

is caused by rainless events during non-vegetal periods. Stated fact has influence on creation of water storage during before-spring periods and enlarges risk of drought.

As an example, 10 longest rainless periods with date of its occurrence, theoretical probability and periodicity of incidence is indicated in **Table 3** for every selection series (year, s0, VP, s0 and VP, s2). The absolutely longest continuous period without rainfall for s0 selection (year, VP) had 35 days and occurred between 25.9.1962 and 29.10.1962. Maximal continuous rainless event extends to 53 days in case of s2 selection for the vegetal period. This event occurred in time interval 19.7.1967–9.9.1967. Precipitation amount was only 1.6 mm in this time interval. This time interval can be considered as the rainless period. From **Table 3** results that the most important rainless periods predominantly occurred in 1960s and after the year 2000.

Extremely dry year 2015 does not appear in s0 selection, however, it was identified in s2 selection. 35 day continuous period during which was total precipitation amount of 2.5 mm occurred. This year was unique in that 91 rainless days between 28.5.2015 and 2.9.2015 were associated in 4 seasons by 18, 22, 16 and 35 days in selection s2.

3.2 Model verification

During the verification of the model GLOBAL the assessment of the weekly values of integral soil water content to the depth of 0.8 m expressed in millimetres of water column was conducted. The selection of the layer for the calculation

Daily precipitation totals to 0,0 mm; period year					Daily precipitation totals to 0,0 mm; vegetation period					Daily precipitation totals to 2,0 mm; vegetation period				
start PWP from	end PWP to	number of days	theoretical value probability	theoretical value periodicity	start PWP from	end PWP to	number of days	theoretical value probability	theoretical value periodicity	start PWP from	end PWP to	number of days	theoretical value probability	theoretical value periodicity
date		[days]	[%]	[years]	date		[days]	[%]	[years]	date		[days]	[%]	[years]
25.09.62	29.10.62	35	1,0	105	25.09.62	29.10.62	35	2,3	43	19.07.67	09.09.67	53	2,1	48
22.10.11	24.11.11	34	1,1	87	18.03.74	14.04.74	28	6,0	17	09.09.06	23.10.06	45	4,5	22
30.01.76	01.03.76	32	2,0	49	05.08.76	31.08.76	27	7,2	14	19.09.62	31.10.62	43	5,4	18
16.02.74	16.03.74	29	3,3	30	14.03.05	09.04.05	27	7,2	14	08.09.61	19.10.61	42	6,3	16
22.10.75	19.11.75	29	3,3	30	15.09.11	06.10.11	22	16,8	6	12.08.74	21.09.74	41	7,1	14
18.03.74	14.04.74	28	3,8	26	24.03.99	12.04.99	20	24,2	4	23.09.00	31.10.00	39	8,8	11
05.08.76	31.08.76	27	4,2	24	23.03.02	11.04.02	20	24,2	4	06.04.62	10.05.62	35	14,3	7
14.03.05	09.04.05	27	4,2	24	02.07.06	20.07.06	19	30,0	3	05.09.89	09.10.89	35	14,3	7
05.10.65	28.10.65	24	6,6	15	14.09.75	01.10.75	18	35,8	3	30.07.15	02.09.15	35	14,3	7
15.09.11	06.10.11	22	9,2	11	17.06.76	04.07.76	18	35,8	3	16.05.64	18.06.64	34	16,0	6

Table 3.
Ten longest rainless periods in every selection series, their theoretical probability of incidence and periodicity.

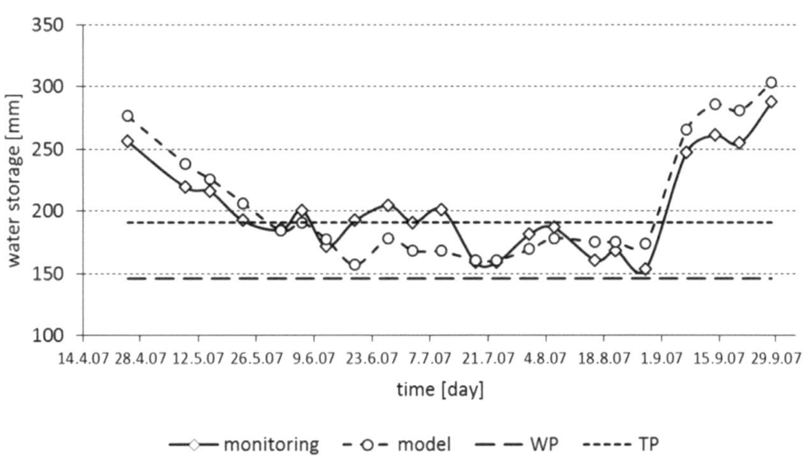

Figure 6.
Comparison between the monitored and the modelled soil water storage up to the depth of 0.80 m during the growing season of 2007.

depended on the depth to which the soil water storage was monitored in field. The results of the simulation and the measurements are shown in **Figure 6**. The diagram shows a very high concordance between the calculated and the measured values which is proven by the results in **Figure 7**. It is a quantile-quantile plot which shows the linear trend and the correlation coefficient between the measured and the modelled values. The R-squared statistic indicates that the model as fitted explains 86.37% of the variability in measurement. The correlation coefficient equals 0.93, indicating a relatively strong relationship between the variables. Other basic characteristics of the descriptive statistics are listed in **Table 4**.

The results show that, especially in terms of moisture, the model GLOBAL $\theta > TP$ tends to overestimate the real state. When the moisture is lower, it tends to underestimate it. In addition, the t-test was applied to compare the mean values and the F-test to compare the standard deviations of the measured and the modelled values. The results of the tests are shown in **Table 5**. The tests showed that the null hypothesis regarding the equality of the mean values and standard deviations of

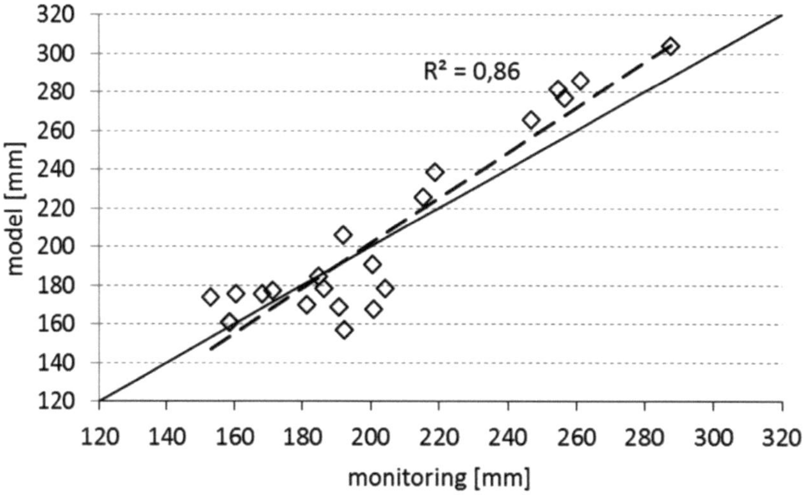

Figure 7.
Representation of the linear dependence via the correlation coefficients between the measured and the modelled daily values of integral soil water content to the depth of 0.8 m in Milhostov.

Statistics	Measurement	Simulation
Count	22	22
Average	202.12	204.43
Standard deviation	37.89	47.84
Variance	1435.57	2288.79
Coefficient of variation	18.75%	23.40%
Minimum	153.32	157.09
Maximum	287.38	303.43
Range	134.07	146.34
Standard skewness	1.455	1.877
Standard kurtosis	−0.284	−0.584

Table 4.
Summary statistics.

t-Test to compare means	
Null hypothesis	Mean measurement = Mean simulation
Alternative hypothesis	Mean measurement ≠ Mean simulation
t-statistic = −0.177316	Two-sided P-value = 0.860112
Conclusion	Do not reject the null hypothesis for alpha = 0.05
F-test to compare standard deviations	
Null hypothesis	Sigma measurement = Sigma simulation
Alternative hypothesis	Sigma measurement ≠ Sigma simulation
F-statistic = 0.627217	Two-sided P-value = 0.293035
Conclusion	Do not reject the null hypothesis for alpha = 0.05

Table 5.
Results of the t-test and F-test.

the measured and simulated values cannot be rejected. The results indicate that the model is suitable for the examined area and it can be used for the simulation of the water regime in the unsaturated soil zone. It should be noted that the model has been verified and successfully applied in other areas around Slovakia (Novák et al., 1998).

Figure 8 shows the contour lines representing the volume moisture and it was made based on the moisture values monitored up to the depth of 0.8 m by 0.10 m thick layers in Milhostov during the growing season of 2007. The picture shows that the whole profile was dry to the subsoil layers.

3.3 Results of numerical simulation

Table 6 lists the basic characteristics of the descriptive statistics applied to the following: seasonal, monthly and daily totals of ET_0, D, P; average soil water storage during the growing season to the depth of 1.0 m and the mean location of GWL under the surface during the growing season. **Table 6** shows that the long-term mean evaporation during the growing season in the form of ET_a is 315.17 mm (59.4% of ET_0) while evaporation ET_0 is 530.63 mm. This leads to the long-term evaporation deficiency of 215.45 mm, which is 40.6% of ET_0. Long-term mean rainfall total during the growing season is 70.0% (371.44 mm) of ET_0. Long-term mean

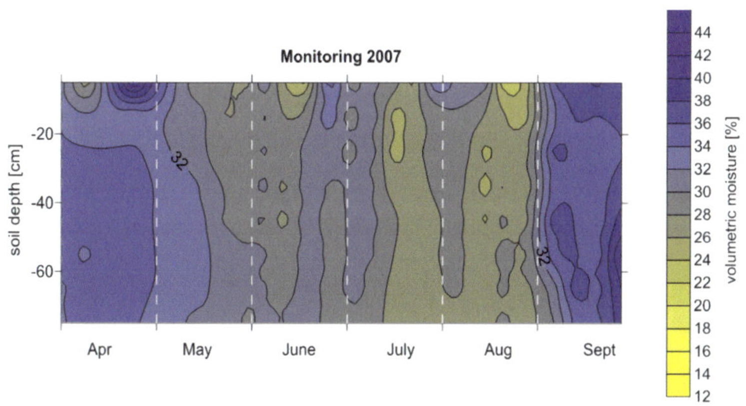

Figure 8.
Contour lines of the volumetric moisture up to 0.80 m during the growing season of 2007.

Statistic	VO						Month		Day	
	ET_0	ET_a	D	WS	P	GWL	ET_0	ET_a	ET_0	ET_a
	[mm]					[cm]	[mm]		[mm]	
Mean	530.6	315.2	215.5	278.4	371.4	135.1	88.4	52.5	2.9	1.7
St. er.	12.0	11.7	16.1	5.5	12.7	3.6	1.5	1.3	0.0	0.0
Median	533.5	290.7	209.0	270.2	360.9	136.2	87.3	51.1	2.8	1.6
St. dev.	81.1	79.5	109.4	37.2	85.9	24.1	25.1	21.9	1.4	1.1
S. var.	6580.3	6316.8	11958.3	1384.9	7373.9	580.3	631.8	477.5	1.9	1.1
Kurtosis	0.5	0.1	1.3	1.2	2.03	3.8	−0.3	0.5	0.1	0.3
Skew.	−0.2	0.7	0.8	0.9	0.93	0.2	0.3	0.5	0.3	0.6
Range	409.9	370.2	556.4	187.0	444.4	159.6	131.9	121.3	9.3	6.0
Min	302.7	156.1	0.2	213.6	226.8	61.1	28.3	3.3	0.0	0.0
Max	712.6	526.2	556.6	400.6	671.2	220.7	160.2	124.5	9.3	6.0
Count	46	46	46	46	46	46	282	282	8235	8235

Table 6.
Statistical characteristics of the seasonal, monthly and daily totals of ET_o, ET_a, D, WS, P and GWL.

monthly total of ET_a is 88.44 mm. Long-term mean daily total of ET_a is 1.73 mm while ET_0 is 2.91 mm. Mean depth of the GWL under the surface is 135.09 cm. **Table 6** shows statistical characteristics of the seasonal, monthly and daily totals of ET_0, ET_a, D, P; mean soil water storage during the growing season to the depth of 1.0 m and mean location of the GWL under the surface during the growing season.

Figure 9 shows the long-term mean values of the monthly total of ET_0, ET_a and D. The results indicate that the highest long-term mean monthly total of evaporation deficiency is in July and August. It is caused by the fact that the mean monthly soil water storage gradually diminishes during the growing season until July when it stops at the minimum value. To the contrary, ET_a increases and reaches the top value in July. The maximum value of long-term mean total of ET_a occurs in June when soil profile contains enough water. From June until the end of the growing season ET_a continuously drops.

Figure 10 shows the development of D, ET_0, ET_a, P totals, mean soil water storage to the depth of 1 m (WS), mean temperatures (T) and mean GWL during the growing seasons between 1970 and 2015.

It is obvious that during the examined period, the difference between ET_0 and ET_a raised. Evaporation deficiency 'D' therefore increases. Variability has also grown during the last 15 years. As for standard deviation 'D', from 1970 to 1985 it was 63 mm, from 1986 to 2000 it was 69 mm and during 2001–2015 it reached 158 mm. The variability has raised in all examined parameters during the last 15 years (e.g. WS raised by 50%). With the exception of evaporation, the trends are balanced.

Table 7 correlates the examined parameters. The most significant and closest relation is between soil water storage and D (**Figure 11**) and GWL which are inversely proportional. On the other hand, water storage in the root zone of a soil profile statistically depends on GWL and ET_a.

The processes can be explained by the fact that GWL is a lower boundary of the unsaturated soil zone and for the purpose of soil water regime, it is defined as the lower boundary condition. The lower boundary of the unsaturated soil zone is thus dynamic and, depending on the interaction with the groundwater, it can change in space and time. When GWL is high, groundwater can reach the soil profile. In depressed lowland areas, such as ESL, groundwater often reaches the surface of the ground and the unsaturated soil zone is lost. Due to its fluctuation

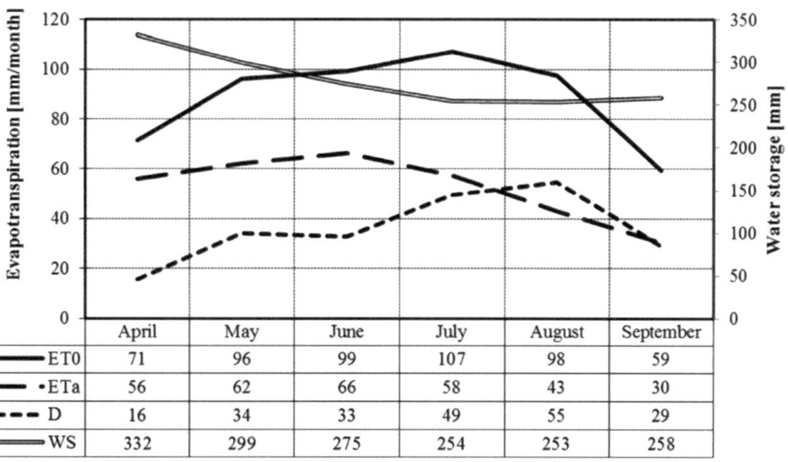

	April	May	June	July	August	September
ET0	71	96	99	107	98	59
ETa	56	62	66	58	43	30
D	16	34	33	49	55	29
WS	332	299	275	254	253	258

Time [month]

Figure 9.
Long-term mean monthly totals of ET_0, ET_a and D.

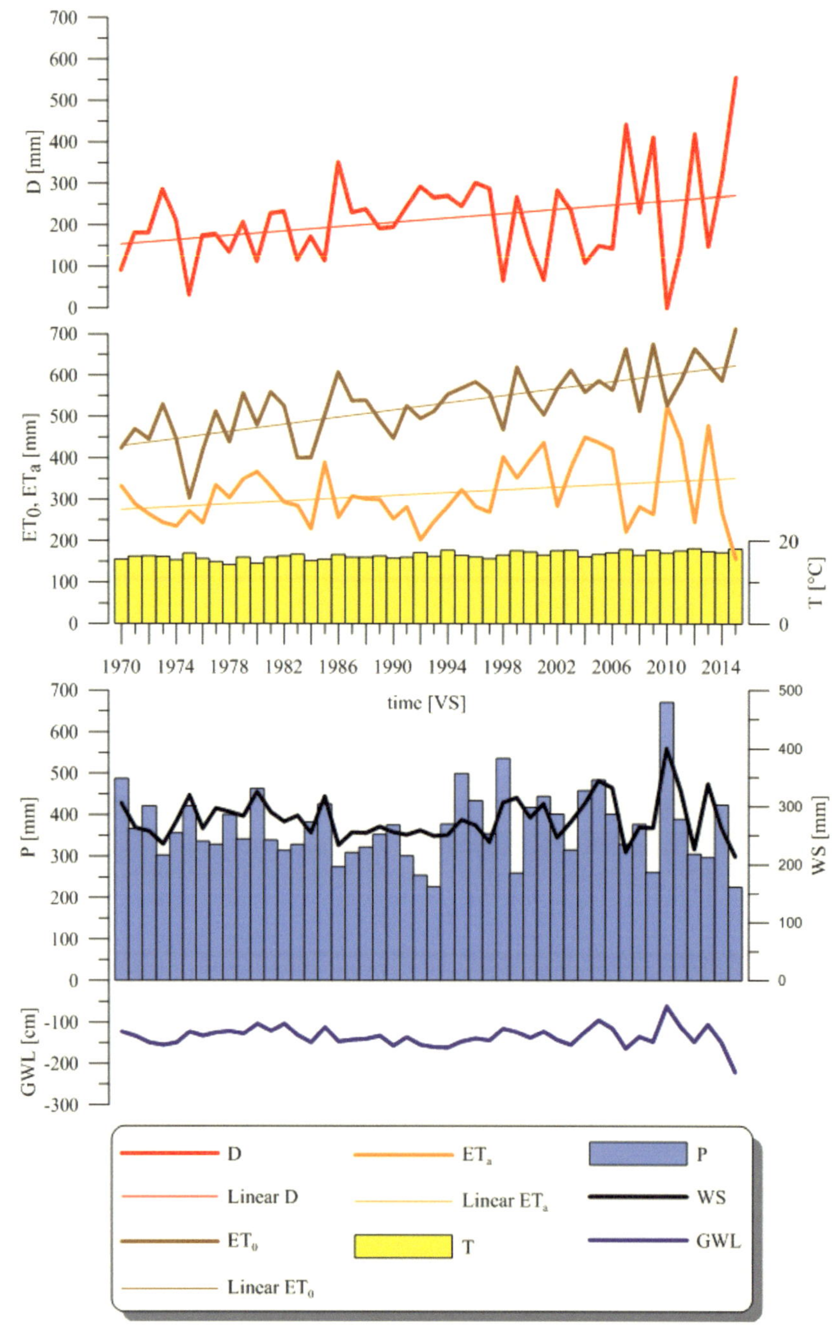

Figure 10.
Evapotranspiration deficiency and the water regime elements during the growing seasons 1970–2015.

in time, groundwater can reach the soil profile even when the average GWL is lower. In this way, the interaction processes influence on the soil water storage and its availability for the plant cover. It is especially observable during periods of meteorological drought. During longer periods with no rainfall, precipitation cannot cover the actual evapotranspiration ET_a and the drying process begins. There is a slight retardation in time in terms of the impact of the drying process on GWL and the unsaturated soil zone. In lowland areas, first layers to be dried are the upper layers of a soil profile. Drying then spreads to the lower layers towards GWL. Time retardation lies in that groundwater supplies the

Parameters	P [mm]	ET_0 [mm]	ET_a [mm]	T [°C]	GWL [cm]	WS [mm]	D [mm]
P [mm]	1.0						
ET_0 [mm]	−0.3	1.0					
ET_a [mm]	0.6	0.1	1.0				
T [°C]	−0.2	0.6	0.0	1.0			
GWL [cm]	0.6	−0.3	0.8	−0.2	1.0		
WS [mm]	0.6	−0.2	0.9	−0.1	0.9	1.0	
D [mm]	−0.7	0.7	−0.7	0.4	−0.8	−0.8	1.0

P—precipitation, ET_0—potential evapotranspiration, ET_a—actual evapotranspiration deficiency, T—mean temperatures, GWL—groundwater level, WS—water storage, D—evaporation.

Table 7.
Correlation table of the examined parameters.

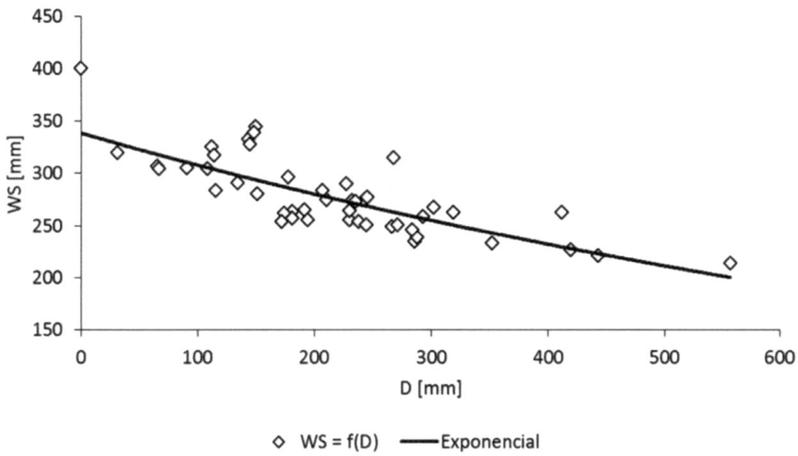

Figure 11.
Graphical representation of the exponential dependence between evapotranspiration deficiency (D) and water storage (WS) in soil to the depth of 1 m; R = 0.8.

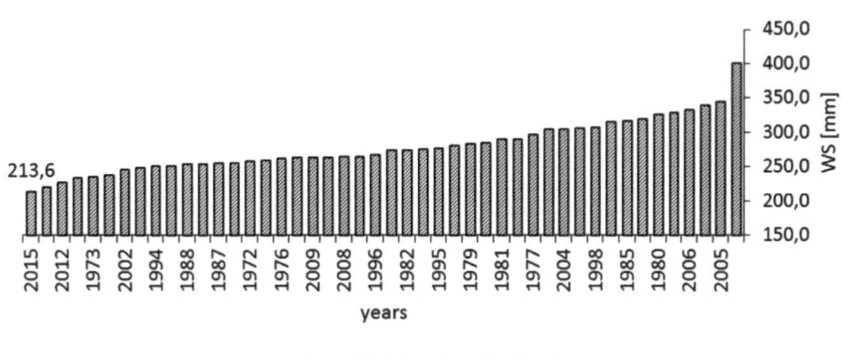

Figure 12.
Growing seasons 1970–2015 ordered by soil water storage in a soil profile to the depth of 1 m.

unsaturated soil zone with water and thus ameliorates its availability for the root zone of the plant cover. In consequence, GWL drops and the unsaturated soil zone becomes thicker. When GWL drops under the critical (threshold) point,

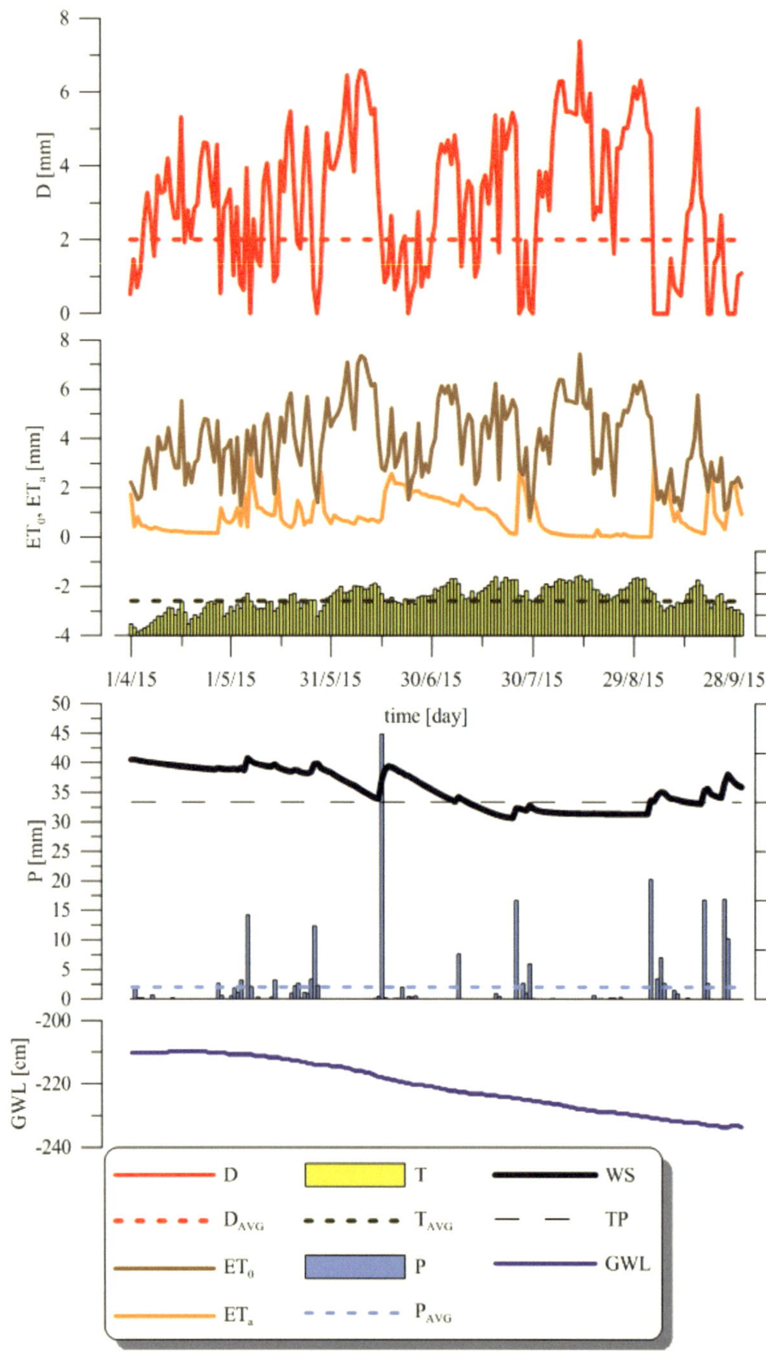

Figure 13.
Evapotranspiration deficiency and water regime elements in the extremely dry year of 2015.

water transfer from GWL to the root zone ceases. Moisture conditions in the balanced layer of the root zone depend solely on rainfall and evaporation. When there is the longer period with no rainfall, water supply towards the roots stops. The upper soil horizons and subsequently the whole root zone get into the state of drought. In **Figure 12** the examined years are ordered by mean soil water storage to the depth of 1 m during the growing season.

The scheme shows that the driest year in terms of soil water storage during a growing season was the year 2015. In consequence, water regime elements in 2015 were analysed. The results of the analysis were calculated with 1-day step and they are shown in **Figure 13**. Evaporation deficiency 'D' is 257% of the long-term mean (1971–2015). ET_0 is 134% and ET_a is 49% of the corresponding long-term mean. The mean temperature is relatively stable, 110% of the corresponding long-term mean. Rainfall 'P' formed 61% of the corresponding long-term mean and water storage in soil to the depth of 1 m 'WS' were 77% of the corresponding long-term mean. The results shown in **Figure 13** indicate that in the other half of the growing season soil water storage dropped under the point of decreased availability.

This corresponds to the fact that the value of 'D' was high above the average. Groundwater continuously drops during the whole growing season. **Figure 13** also shows the development and correlation between ET_0, ET_a and D.

4. Conclusions

Precipitation amount and temporal distribution of the rainfall is important for water refilling of the environment for balanced periods. Drying of soil profile occurs during long rainless periods. Meteorological drought and subsequently soil drought occurs in the case of the sufficiently long rainless period. Therefore it is necessary to know size and statistical characteristics of rainless periods (RLP). The aim of the contribution is to identify important rainless periods, quantify temporal lengths, probability characteristics and trends of RLP. Climatic station of Milhostov (N 48°39,786′; E 21°43,298′) was chosen for selection of statistically important rainless periods. Station represents wider area of lowland. Daily precipitation amounts of the 1961–2015 period was examined for the station. 20,080 daily precipitation amounts (including zero) were analysed for the period. Length of RLP was identified in two ways. Periods with zero daily precipitation amounts were considered in first selection (s0). Daily precipitation amounts lower than 2 mm were considered as zero in second selection (s2). The absolutely longest continuous period without rainfall for s0 selection (year, VP) had 35 days and occurred between 25.9.1962 and 29.10.1962. Maximal continuous rainless event extends to 53 days in case of s2 selection for the vegetal period. This event occurred in time interval 19.7.1967–9.9.1967. Number of rainy and rainless days is for individual years and their vegetal periods balanced in the long term view. Only duration of rainless events of non-vegetal periods changes in examined periods. Growth of time length of rainless days duration was identified in non-vegetal half-year.

The study analysed the development of ET_0, ET_a, D, WS, P, GWL location and T during the growing seasons during the years 1970–2015 on the basis of the measured data and the data gained via numerical simulation. The interaction processes in the root zone of a soil profile during the creation of soil water regime were quantified. The quantification is crucial for understanding the processes occurring during drought creation, duration and termination. It has been demonstrated that soil water storage depends heavily on evaporation, i.e. actual evapotranspiration ET_a. Subsequently, actual evapotranspiration influences on evapotranspiration deficiency 'D' and on the location of GWL. For that reason, evapotranspiration deficiency 'D' can be considered an indicator of the drying of a soil profile. Drying starts when water inflow towards plant roots is reduced. During the state of evapotranspiration deficiency, groundwater supplies the root zone of a soil profile with water. When there is the long period with no rainfall, water transfer from GWL towards plant roots ceases. Moisture conditions in the balanced layer of the root zone depend solely on rainfall and evaporation. The processes are demonstrated on the driest growing season of 2015. From the point of view of soil water storage to the depth of 1 m this season was absolutely driest growing season in the examined period from 1970 to 2015.

Acknowledgements

This work was created thanks to financial support of VEGA project 2/0062/16.

Author details

Milan Gomboš*, Branislav Kandra, Andrej Tall and Dana Pavelková
Institute of Hydrology, Slovak Academy of Sciences, Bratislava, Slovak Republic

*Address all correspondence to: gombos@uh.savba

IntechOpen

© 2019 The Author(s). Licensee IntechOpen. This chapter is distributed under the terms of the Creative Commons Attribution License (http://creativecommons.org/licenses/by/3.0), which permits unrestricted use, distribution, and reproduction in any medium, provided the original work is properly cited. (cc) BY

References

[1] Van Genuchten M. A closed-form equation for predicting the hydraulic conductivity of unsaturated soils. In Soil Science Society of America Journal. 1980;**48**:892-898

[2] Majerčák J, Novák V. Simulation of the soil-water dynamics in the root zone during the vegetation period. I. The mathematical model. Vodohosp. Čas. 1992;**40**(3):299-315

[3] Novák V, Majerčák J. Simulation of the soil-water dynamics in the root zone during the vegetation period. II. The course of state variables of soil water below the maize canopy. Vodohosp. Čas. 1992;**40**(4):380-397

[4] Majerčák J, Novák V. GLOBAL, one dimensional variable saturated flow model, including root water uptake, evapotranspiration structure, corn yield, interception of precipitations and winter regime calculation. In Research Report, Bratislava, Institute of Hydrology, Slovak Academy of Sciences. 1994, 75 pp

[5] Šimůnek J, Huang K, Šejna M, Van Genuchten Th M, Majerčák J, Novák V, Šútor J. The HYDRUS - ET Software Package for Simulating the One - Dimensional Movement of Water, Heat and Multiple Solutes in Variably - Saturated Media. Version 1.1.1997, Institute of Hydrology S.A.S. Bratislava - U.S. Salinity Laboratory, Riverside, 1997, 184 pp

[6] Novák V, Šútor J, Majerčák J, Šimůnek J, Van Genuchten M Th. Modeling of Water and Solute Movement in the Unsaturated Zone of the Žitný Ostrov Region, Institute of Hydrology S.A.S. Bratislava, 1998 73 p

[7] Šoltész A, Baroková D. Analysis, prognosis and design of control measures of Ground water level regime using numerical modelling, Podzemná voda, XII, SAH, Bratislava 2006;č.2:113-123

[8] Šoltész A, Baroková D. Impact of landscape and water management in Slovak part of the Medzibodrožie region on groundwater level regime. In Journal of Landscape Management. 2011;**2**(2):41-45

[9] Červeňanská M, Janík A, Šoltész A, Baroková D, Gramblička M. Determination of input data for utilization of drainage channel systems on improvement of water regime in wetland systems. In Acta Hydrologica Slovaca. 2016;**17**(2):133-139

[10] Estokova A, Balintova M. 2018. Advances in Environmental Engineering. In ENVIRONMENTS. ISSN 2076-3298. 2018:**5**(5)

[11] Hlaváčiková H, Novák V, Kostka, Z, Danko M, Hlavčo J. The influence of stony soil properties on water dynamics modeled by the HYDRUS model. In Journal of Hydrology and Hydromechanics, 2018;**66**(2):181-188

[12] Igaz D, Šimanský V, Horák J, Kondrlová E, Domanová J, Rodný M, Buchkina N. Can a single dose of biochar affect selected soil physical and chemical characteristics. In Journal of Hydrology and Hydromechanics, 2018;**66**(4):421-428. ISSN 0042-790X

Chapter 6

Striking Challenges on Water Resources of Lebanon

Amin Shaban

Abstract

Lebanon was described as the "water tower of the Middle East," but the current status is contradictory, and water supply/demand gap has become imbalanced. Even though, water is available, yet water shortage is a major challenge in Lebanon. This is attributed to the undefined hydrological cycle and the lack of sufficient hydrologic data to establish proper strategies and policies. This study investigates the existing physical and anthropogenic challenges and their impact on water resources. Hence, climate change is one major aspect of these challenges where the precipitation patterns have been abruptly changes to torrential rainfall, and this is accompanied with increased temperature estimated at 1.6°C. In addition, the exacerbated population rate (2%) with the doubled number of refugees makes Lebanon a country under water stress. The discharge in rivers and springs has declined by about 60% over the last four decades, while groundwater level has sharply lowered in the major aquifers and the pumping rate has decreased to about 35%. The problem on water resources in Lebanon should be resolved by promoting a national strategy that takes all hydrological components into account, as well as new adaptation measures must be proposed for better water management.

Keywords: water deficit, river discharge, pollution, mismanagement, water policy, Lebanon

1. Introduction

The topography of Lebanon is characterized mainly by mountainous ridges where two mountain chains extend parallel to the Mediterranean Sea, thus building a regional climatic barrier besides the cold air masses from the sea, and then falling considerable amounts of rainfall and snow in the elevated areas. The annual rainfall rate ranges between 700 and 1500 mm and snow remains for a couple of months on mountains and covers more than 2500 km^2 of the Lebanese territory. This is reflected on the hydrology of Lebanon where 12 permanent watercourses (i.e., small rivers) and more than 2000 major springs (>10 *l*/sec) occur. Moreover, there are a number of aquiferous formations and karstic conduits where considerable amount of groundwater is stored. This makes Lebanon a country with sufficient water resource. This status remained until the beginning of 1990s when the water sector in Lebanon started facing severe geo-environmental problems regarding water volume and quality.

Recently, water demand/supply has occupied a wide space of argument in Lebanon, and this has been raised mainly on the national level, notably that water per capita has been decreased to more than 60%. Hence, it is a paradox that

IntechOpen

Lebanon became one of the countries under water stress since it was ranked as 149 on the wide world water availability list.

There are many challenges existing in the water sector of Lebanon, including physical and anthropogenic aspects. These have been exacerbated lately and their impact on water supply has been reaching intolerable levels. The majority of these challenges imply a change in climate, and more certainly increased temperatures and oscillating rainfall patterns. In addition, there is the population growth which has added to water demands as well as the issue of refugees who comprise more than 40% of Lebanon's population.

Complaining for water becomes a daily issue between different concerned institutions, and there is increased debate on the responsibility for the deterioration in the water sector of Lebanon, notably that the severe situation reached all surface and subsurface water resources and covering all Lebanese regions. Thus, every major water resource has come under dialog; and, many campaigns, committees, and even institutional frameworks were established to meet water demands, but no enhancement has been observed to date.

The Litani River, the largest of its type in Lebanon, is a typical example for the unfavorable status of water sector in Lebanon. The river is witnessing an extremely severe situation, and the streamflow has become minimal while the quality has ultimately deteriorated and exceeded hundreds of times the international norms. According to Shaban and Hamze, the river's situation was described as "Death of a River" [1].

For the Litani River, a business plan to identify the measures to alleviate pollution in the river and the belonging Qaraoun Reservoir, was framed in 2006 and followed by an updated plan in 2013. Meanwhile, a national multi-ministerial committee for the depollution of the river basin was also established in 2012, and consequently a national committee was created to follow up the implementations for the remediation of the Qaraoun Reservoir in 2014 [2]. In addition, loans from international entities (e.g., World Bank) were provided to support the implementation of the approved plans. Nevertheless, the river is still in its worst condition.

This chapter aims at clarifying the major existing challenges facing water resources of Lebanon whether these challenges are of the natural or anthropogenic aspects. This will help highlighting the fundamental elements of these challenges, which can, further, serve as clues for proposing management solutions. The analysis of these challenges is based on previous studies obtained by the author who used several tools for analysis including, mainly, hydrological tools, as well as remote sensing and geo-information systems.

2. Method of analysis

For studying the challenges of water needs to investigate the existing problems on water supply/demand, the main factors influencing water supply to consumers must be identified and then investigated accordingly. This can be elaborated by considering the two principal pillars of challenges in the water sector. The challenges are of time-related origin, and they can be classified as: (1) challenges that originally occurred long time ago in the past, and they belong to the physical origin of the area and the living people inside and (2) challenges that have recently interfered with and influenced water volume and quality. For both aspects of time-related origin, the method of analysis of water challenges can be elaborated by empirical assessment of the water resources under stress, and this can help to identify what actions should be taken to reduce the impact of these challenges.

According to Shaban [3], the principal phases for the methodology used in analyzing challenges in the water sector are illustrated in **Table 1**. They are five major phases

Phase	Description	Expected outcomes
Data and information inventory	• Previous studies, researches, etc. • Implemented plans and projects • Existing strategies and policies • Long-term climatic and hydrologic records	Preparing inventory on water resources in the country
Data analysis and interpretation	• Map analysis (including illustrators) • Obtained case studies on water resources • Statistical analysis of measures (e.g., water consumption population size, etc.)	Identifying the current measures and their trends
Use of advanced tools	• *In-situ* measurements • Advanced laboratory techniques • Remote sensing and geo-spatial products	New tools and techniques will give accurate results
Field survey	• Field investigations (mainly those obtained by the author) • Results of data from previous field observations and investigations	Verifying obtained results and adding filed measures
Benefit from lessons learned	• Benefit from successful studies and projects obtained regionally and globally • Identifying useful tools used in water resources assessment • Inducing reasons of failure in water sectors	Determining elements of success or failure in water sector and benefit from these element for better management approaches

Table 1.
Major phases of analyzing challenges on water resources (Adapted from Shaban [3]).

that are integrating different aspects of data and information where the trends on water resources are viewed over long time period from past to recent. For each phase of data analysis, there are a number of measures and information sought (**Table 1**). In this respect, the availability of data and records is significant for the analysis.

2.1 Data sources

Constraints always exist while collecting data and information for further analysis in hydrological and meteorological studies. This is well pronounced in Lebanon where there is a lack of continuous time series data on elements of water assessment, which mainly includes the precipitation rate, discharge from rivers and springs, as well as volumetric measures on groundwater. There is also obvious contradiction between the measures obtained on water resources, and this is attributed to data unavailability, non-uniform distribution of measuring stations (meteorological and hydrological), and the complicated geomorphology of Lebanon.

For this research, the following data series was prepared:

- Precipitation and temperature datasets for the last four decades.
- Water flow rate for the major rivers and springs, as well as the two main artificial lakes in Lebanon.
- Water table depth and pumping rate from the major two aquifers in Lebanon and for selected areas.
- Results of water consumption surveys from pilot areas in Lebanon.

The collected data in this research were primarily organized; therefore, gaps were identified and filled either by statistical interpolation or from the remotely sensed data.

The majority of climate and hydrologic data sources, which are necessary for the assessment and for identifying the existing challenges, can be summarized as follows:

- Climatic Atlas of Lebanon—Atlas Climatique du Liban [4, 5].
- Tropical Rainfall Mapping Mission [6], which is a remotely sensed system based on radar data. It retrieves rainfall datasets on a daily basis.
- Department of Irrigation and Agro-meteorology (DIAM), Lebanese Agricultural Research Institute [7].
- Climate Hazards Group InfraRed Precipitation with Stations (CHIRPS) dataset. Its algorithm is built on ground-climatic measures incorporated with satellite information. It provides daily and monthly data [8].
- General Directorate of Civil Aviation—Direction Générale de l'Aviation Civile [9].
- NOAA climatic data system—National Oceanographic Data Center [10].
- Data and measures found in studies (e.g., theses, technical reports, etc.) conducted by different authors, all of them for pilot areas from Lebanon.
- Data and information measured in filed surveys whether on water resources or on the socioeconomic status, with a special emphasis on water consumption.
- Data from the projections and scenarios elaborated for the hydro-climate of Lebanon, such as those obtained by [11–13]

Thematic data were also used in this research in diagnosing the physical influencers on water resources. These mainly include a number of thematic maps and illustrations, such as topographic, geologic, hydrological, and hydrogeological maps with their attribute measures. They could be used for analyzing major water-related parameters.

2.2 Data analysis

The majority of data analysis follows a simplified approach for data investigation, comparison, and deduction. This represents a synthesis of the existing challenges on water resources. The followed approach is based on the data collected from all mentioned sources. Therefore, the method of analysis is as follows (**Figure 1**):

1. Review of the available documents and records on water resources and the related influencing factors (e.g., climate), and this focuses mainly on the volume and quality of the major water resources in Lebanon. Hence, notes were taken on the fundamental findings with a special emphasis on:

    - Water inputs into the hydrological systems (e.g., watersheds, aquifers, etc.).
    - Aspects of water loss (e.g., flow into the sea, evapotranspiration, etc.).

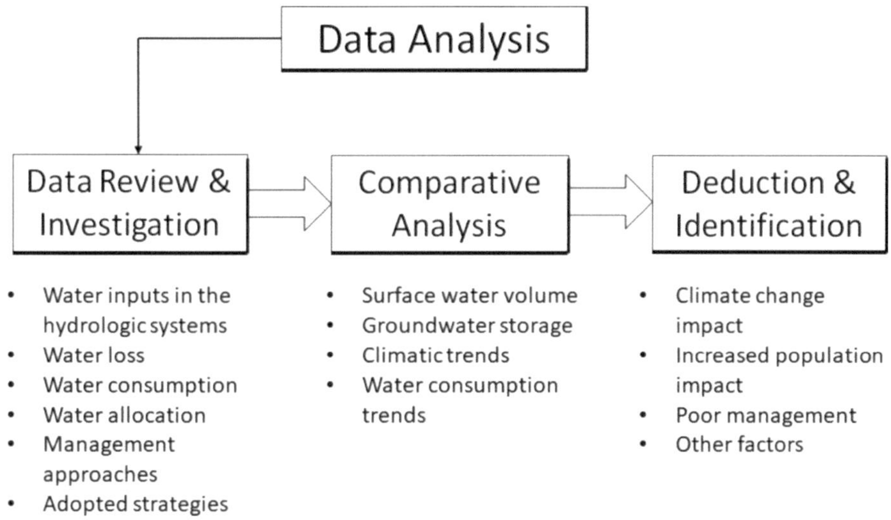

Figure 1.
Major items for the method of analysis.

- Water consumption by sector and allocation for different purposes.

- Existing aspects of water exploitation, notably for irrigation.

- The obtained management approaches whether taken by individuals or the government sector.

- The adopted strategies and plans for water conservation and management.

2. Applying comparative analysis on different water measures based on time series data analysis. This implies the use of advanced statistical methods to identify the behavior of the numerical values, which are mainly of hydrological and meteorological origin in addition to the consumption rates. In particular, singular spectrum analysis (SSA) and Fisher-Shannon and detrending fluctuation analysis (DFA) methods were applied to:

    - Water volume in rivers, springs, and lakes (artificial and man-made).

    - Groundwater depth, discharge, and number of boreholes.

    - Climatic trends mainly including the precipitation and temperature.

    - The changing trend in water consumption and water supply/demand over long times series.

3. Identifying major clues toward understanding existing challenges. In other words, it implies determining the influencing factors that result in challenges on water resources of Lebanon. This highlights:

    - The impact of climate changes (considering space and time dimensions).

    - The degree of influence caused by the increased population and refugees.

- Investigating the influence of poor management on water resources.
- Searching all other factors that directly or indirectly act on creating challenges on water resources.

3. Physical factors

Debate always occurs when talking about the responsibility of the existing challenges on water resources and the question is often raised: Are the existing challenges on water resources physical or man-made? Actually, both types of challenges occur in Lebanon and nobody has identified which is the most effective one, and even the degree of influence of each challenge is still obscure. However, there are many clear evidences to highlight the factors behind these challenges, and these evidences are declared even by the individuals who are directly touched by water deficit in all regions of Lebanon.

There are several natural factors that act on the hydrological regime of water resources. These are of different origins and act at different scales. They result in challenges that are represented mainly by shortage in water supplied to the consumers. Thus, some of these factors exist alone and sometimes are found combined with each other. Hence, physical factors are well known in Lebanon and they influence water regime from the source, through the entire water cycle including storage, consumption and loss into the sea.

The following are the major physical factors that act in developing challenges on water resources of Lebanon.

3.1 Geo-hydrological factors

The geo-hydrological factors have a significant role in creating and enhancing the impact of challenges on water resources of Lebanon, notably that Lebanon's territory encompasses all processes and features of the hydrologic cycle. This is highly influenced by the distinguished geomorphology of Lebanon, which possesses mountain ridges with sharp peaks that are detached by several valley systems and then carrying surface water directly to the sea.

The geo-hydrological setting of Lebanon implies different aspects of the elements influencing water flow/storage regime where the majority of this regime results in water seepages, inaccessibility, and water loss. This unfavorable flow/storage regime acts negatively on the exploitation of water resources whether those on surface or the subsurface. Therefore, the influencing elements on water resources in Lebanon include the following:

1. Lithological characteristics: The rock types that form the stratigraphic succession of Lebanon are characterized by diverse lithological aspects. This includes interbedded and intervening permeable and impermeable rock strata with relatively thin bedding planes, and this in turn enhances the seepage of water at the contact between the two diverse lithological types. This geo-hydrological phenomenon is well pronounced between the carbonate (e.g., limestone and dolomite) and argillaceous rocks in Lebanon.

2. Rock deformations: This is represented mainly by the dominant fracture systems, which characterize the rocks of Lebanon. Thus, Lebanon is known by its complicated and active tectonic setting along the Dead Sea Rift System; therefore, it is dominated by rock deformations mainly including dense fracture

systems (e.g., faults, joints, and fractures) that cut among different geological rock formations, as well as foldings of rock strata. Both aspects of rock deformations induce sharp bedding inclinations. The existing rock deformations often play a negative role when: (a) they transit groundwater from the coastal aquifers into the sea, and then water loss occurs [14, 15]; (b) groundwater seeps into very deep (unreachable) aquifers; and (c) groundwater leaks from the major aquifers into undefined systems [3].

3. Acute sloping terrain: It is one of the most acting factor on surface water flow regime, notably that the rugged topography of Lebanon makes it a terrain with less water retention. Therefore, the complicated geology of Lebanon is reflected by its topography and existence of acute slopes. In this respect, the slope gradient is often high and the average slope gradient exceeds 25 m/km over most of the Lebanese mountain chains (**Figure 2**). The steep sloping terrain negatively acts in enhancing water flow energy which can be to the sea or the neighboring countries and creates water loss before any proper investment for managing these resources.

   Also, the rapid water flow along the incident valley courses will act in retarding the percolation rate of surface water into substratum, and thus large amounts of surface water is lost before sufficient time for natural recharge to groundwater reservoirs [16]. However, such challenge can be managed if surface water restrictions (i.e., harvesting) are built, such as dams, channels, lakes, etc. [3].

4. Karstification: This is a very common hydrological phenomenon occurring in Lebanon and it is often given concern, notably because karstification has a significant influence on groundwater flow/storage regime though the subsurface routes, galleries, cavities, and shafts. The karstification features have different dimensions and occur on a large scale, and thus they are able to transport water for several kilometers vertically and laterally. Also, karstification features often

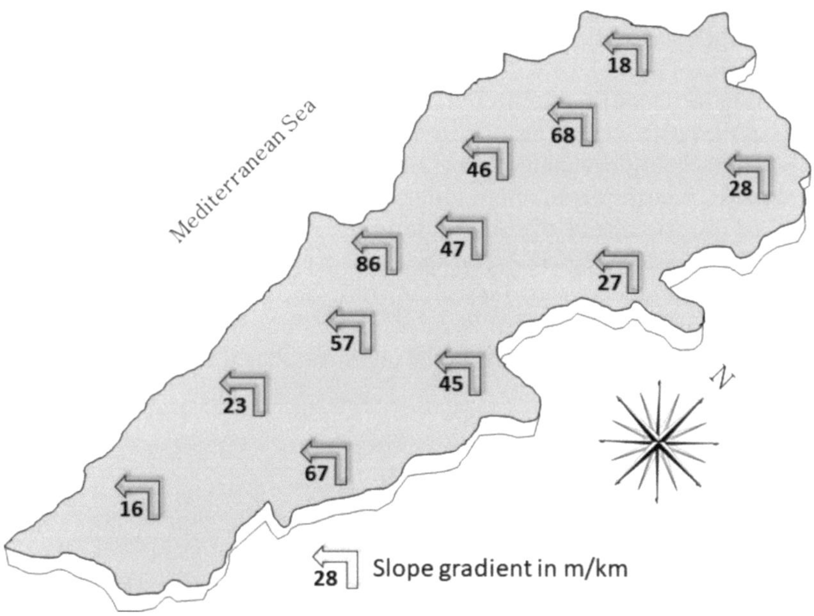

Figure 2.
Simplified map showing selected slope gradients in Lebanon.

play similar roles in transporting groundwater from the coastal aquifers into the sea (representing invisible water routes) or into deep and undefined aquifers [16].

Being of utmost importance to subsurface features, the karstic conduits and galleries are often found with large amounts of groundwater (e.g., Jeitta grotto), but there are many other aspects of conduits that are supposed to exist in Lebanon, and they are still uncovered. In fact, the karstic conduits usually form complicated hydrological systems, especially when exploration methods for groundwater are applied. Thus, these conduits do not have uniform or defined alignment, which makes them difficult to be identified. In addition, these conduits create problems while proposing artificial recharge into the coastal aquifers of Lebanon.

3.2 Meteorological factors

The meteorological factors have a significant role in creating challenges on water resources, and many people consider that the changing climate is the major reason behind water deficit in Lebanon. As a matter of fact, studies on this respect are tremendous, but they have contradicted results because of the lack of measuring records.

Generally, Lebanon is characterized by sub-humid to humid climate, but the country is still witnessing significant water shortage and imbalanced supply/demand. The annual rainfall rate in Lebanon ranges between 650 mm on the coast and 1500 mm on the adjacent mountains, while the average is about 910 mm. In addition, snow covers more than 2500 km^2 (25% of the Lebanese territory) and remains on mountain crests for a couple of months, while the average temperature in Lebanon changes between 21 and 23°C.

Over the last four decades, Lebanon has been under oscillating climatic conditions and this has reflected in its water availability; and, it can be said that the climatic oscillations play a role in determining the hydrological regime of many surface and sub-surface water resources, but this is not the only reason behind the water problem in Lebanon. The discharge in rivers and springs has decreased by about 60% over the last four decades. In addition, groundwater showed abrupt decrease in the pumped rate and water table has lowered in the main aquifers.

According to the most recent climatic data analysis, results showed that there are obvious periodicities in the climate regime of Lebanon, which is accompanied with meteorological oscillations [17], and the estimated rainfall rate was found to be fluctuating from one region to another in the range of ±200 mm over the last four decades [18], but no noticeable change has been reported in the volume of rainfall (**Figure 3**). However, there is a changing rainfall pattern and rain has been more torrential; in addition, there is clear shift in the beginning and ending of seasons. Besides, an obvious increase in the average temperatures has been reported over the last four decades (**Figure 4**), and it was calculated at about 1.8°C.

3.3 Water loss

Loss of water without any proper exploitation is one major problem that challenges the Lebanese water sector. Thus, many aspects of water loss occur in Lebanon, mainly including the seepage of groundwater to the sea along rock deformations and karstic conduits, flow of groundwater into deep karstic aquifers,

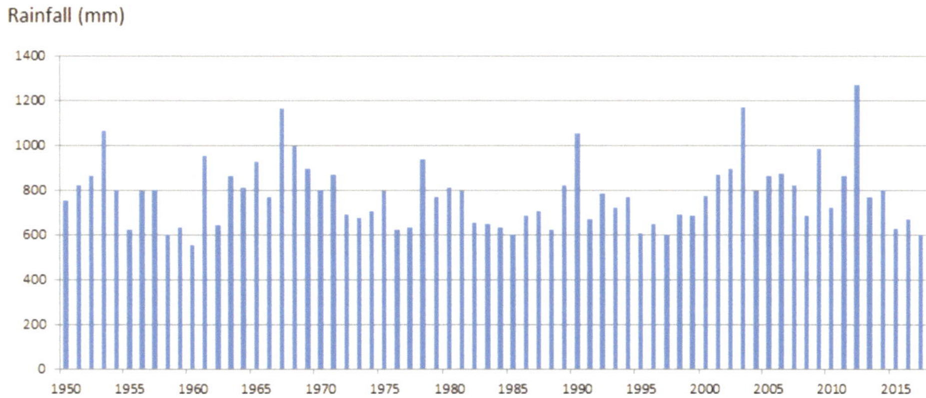

Figure 3.
Annual average rainfall rate in Lebanon [18].

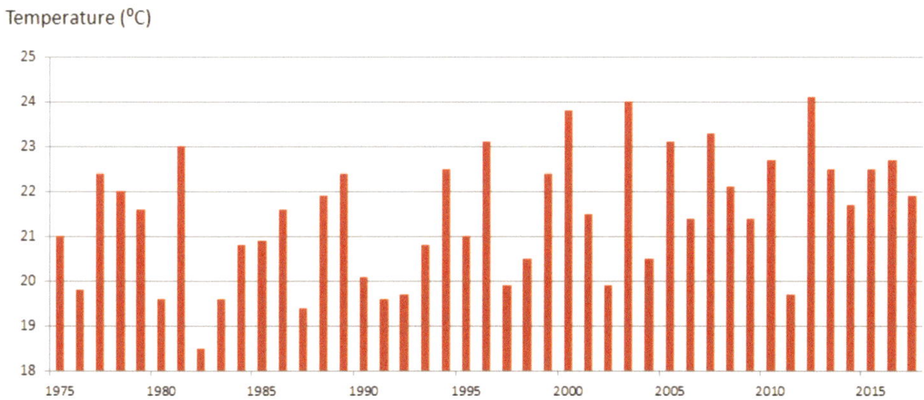

Figure 4.
Annual average temperature in Lebanon [18].

and the relatively high evapotranspiration rate (>55%). Also, there is a large part of surface water and groundwater in Lebanon shared with neighboring countries. Thus, of the 803-km perimeter of Lebanon's territory, about 559 km (67%) is shared with Syria in the north and east, and 9 km (11%) is shared with the Palestinian territory in the south and partly in the southeast.

About 27.5% of Lebanon's area is shared as transboundary river basins with the riparian regions, where Lebanon comprises one of the major tributaries for two international rivers; these are the Orontes and Jordan rivers (**Figure 5**). Therefore, water flows from the transboundary rivers of Lebanon's rivers with no/or very little utilization, while the estimated annual discharge from these rivers (from the Lebanese side) is about 867 million m^3 [17].

Shared groundwater with the neighboring regions for Lebanon occupies about 2631 km^2, which is equivalent to 25% of the Lebanese territory. The sharing reservoirs represent potential rock formations for groundwater storage. However, the geological setting of Lebanon makes groundwater flow along subsurface routes mainly from Lebanon to the neighboring regions. According to the applied hydrological calculations, the shared aquifers in Lebanon contain approximately 365 million m^3 of water [17].

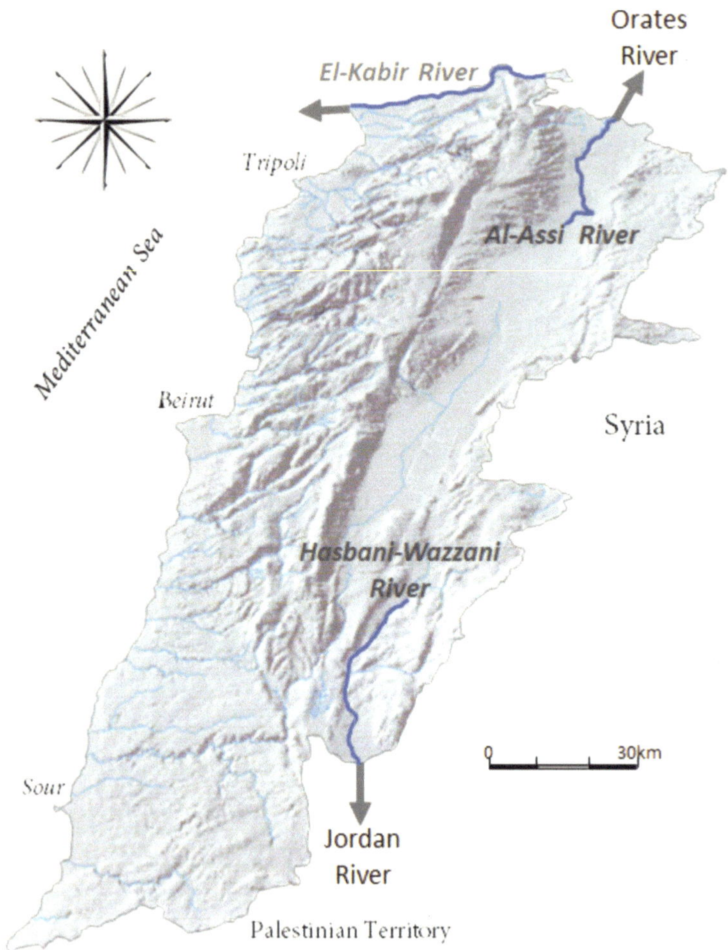

Figure 5.
Transboundary rivers of Lebanon.

4. Anthropogenic factors

There are several anthropogenic factors that play a significant role in increasing the impact of challenges on water resources, and these factors occur in Lebanon and remained untreated. However, the majority of anthropogenic challenges on water resources are represented mainly by the increased population and the related increase in water demands. This also results in many aspects of water deterioration (quantity and quality) and water loss due to the unwise use of these resources. Nevertheless, the anthropogenic factors that influence water resources can be resolved if consumers regulate their use of water and this in turn needs awareness and economic controls on water use.

4.1 Increased population

Normally, population size increases, but it would be wisdom to adapt the water use with the existing population size. In other words, the ratio of increase in population should be accompanied with more management approaches to have steady state of supply/demand. However, this is not the case in Lebanon where the increased population adds more stress on water supply.

The population growth rate in Lebanon is estimated at 1.1% [19]. This means that there is about 1.1% annual increase in demand for water and this also will be joined by additional human requirements. Of course, this will result in more water demand for different purposes, which is estimated at 12.8 million m^3/year for the whole country.

According to the estimated water demand in Lebanon, which is about 220 m^3/year/capita [20], this estimate will increase by about 2.5 m^3/year/capita. **Figure 6** reveals an example of the increased water demand integrated with the population density.

4.2 Unwise use of water resources

Due to the lack of awareness for wise-use of water resources in Lebanon, the existed water supply deficit often results negative behaviours of consumers. This is well pronounced in the agricultural sector (which accounts for about 70% of water use) where farmers often follow chaotic irrigation approaches. Therefore, uncontrolled digging for boreholes followed with extensive pumping has become a common observation. Pumping of water directly form rivers and springs is also widespread, and this has reflected in the loss of volume of water in surface water bodies and in groundwater reservoirs.

The unwise use of water resources in Lebanon accompanied with the lack of governmental controls have reflected in poor water quality and quantity, and thus several examples on water deterioration have been witnessed. The Litani River is one example where the water quality has been totally deteriorated and water volume has become minimal and therefore, the river is in jeopardy.

4.3 Lack of governmental controls

There is still a debate on the management status of water sector in Lebanon, notably that the country has witnessed several geo-political conflicts that have resulted in negative impact on water resources and the related sectors (e.g., agriculture, energy, and food). These conflicts have created distortion in most of the gauge stations (hydrological and meteorological), break down in the institutional framework, and ignorance of legislations to mitigate the encroachments on water resources, plus the lack of adaptation instruments to face the physical and man-made

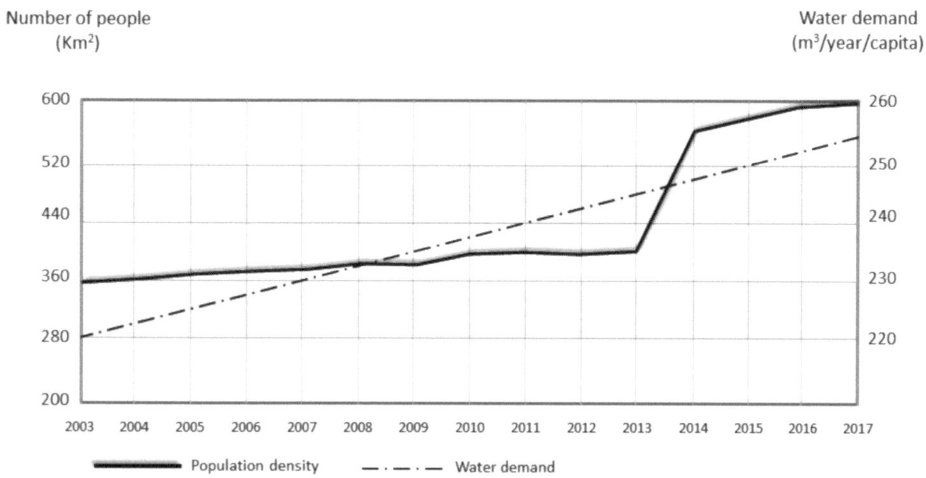

Figure 6.
Population density and water demand in Lebanon between 2003 and 2017.

changes, and certainly change in climate and new water demands. The unfavorable situation has made it difficult to implement any economic policy in order to regulate water consumption and collect financial resource to develop the water sector.

5. Conclusion and discussion

It is a paradox that the estimated water availability in Lebanon is about 1350 m^3/year, while the demand does not exceed 220 m^3/capita/year. Besides, water supplied by the government sector does not exceed 35–40% of the total demand, which is equivalent to about 83 m^3/capita/year [21]. The rest needed amount of water is often managed by the consumers, and this adds more financial commitments notably for the people with limited income.

The country is witnessing one of the most severe conditions in the water sector and challenges on water resources have been exacerbated day after day. The majority of challenges include: limited water supply and water impurity. Therefore, people in Lebanon are always complaining about water shortage as well as its quality.

Based on the analyzed data, the following is a summary on the challenges of water resources in Lebanon.

1. Challenges on water volume:

    - The decreased volume of discharge from rivers and springs (about 55–60%) and the lowering of lakes' capacity (up to about 65%) have resulted in shortage in water supply, which is estimated at 50–55%. This has obligated consumers to look for new supply sources (i.e., buying water) costing them 120–150$/month on average.

    - Groundwater volume in the major aquifers of Lebanon has declined by 35–40%. This implies lowering of water table (30–35 m) and decrease in its pumping rate. This makes it necessary to increase the depth of drilled boreholes in order to attain sufficient amount of water. Therefore, a cost of about 1500–2000$ has been added.

    - Most wetlands in Lebanon are dried or have lost their unique ecosystems to water deficit and climatic variability. This leads to loss of the economic value of these lands, especially most of them have become abandoned.

    - Water supply by the government sector is less than 35% of water demand and this has resulted in several financial and socioeconomic problems.

    - The decreased water level in the Qaraoun Reservoir, the largest artificial water body in Lebanon, by about 45–48% has affected the hydropower generation from the reservoir, estimated at 238 megawatt, which is equivalent to about 10.5% of electricity needs of Lebanon [1].

2. Challenges on water quality:

    - It can be said that water from all rivers in Lebanon is polluted. For example, heavy metals content is very high in the river where Cr is 5 times above the norms and Cu level 800 times than the norms [22]. The existing pollution makes it difficult to use river water for domestic use, and this was also compensated by looking for alternative sources of pure water.

- Groundwater contamination (biologicallyor chemically) is well pronounced, notably in shallow aquifers. Thus, most people fixed water filters, which has been lately become commercial issue.

- Surface water bodies are also contaminated. The Qaraoun Reservoir, the largest of its type in Lebanon, is witnessing severe quality deterioration, and thus toxic bacteria have been reported in the Qaraoun Reservoir [23].

3. Challenges in water-agricultural-food nexus:

    - There is a change in agricultural practices, mainly including planting of new crops that are tolerant to dryness, and this resulted problems in the agronomical marketing. The cost of many filed crops has been lately doubled.

    - Insufficient water motivated farmers to use excessive amounts of fertilizers and chemicals and this in turn added contamination to the shallow groundwater and then to crops. This influences the food security in Lebanon since 10–15% of these crops contain heavy metals [24].

    - Water deficit for irrigation, by about 30–40%, created unfavorable socioeconomic and demographic changes from the rural to urbanized areas, notably to the coastal zone which comprises more than 70% of Lebanon's population.

    - Insufficient water supply postponed several hydro-power stations in Lebanon.

Based on the existing challenges, however, the following recommendations can be suggested:

1. Awareness and advises for better use of water should be introduced to consumers.

2. The existing strategies (i.e., The National Water Strategy of Lebanon) and policies should be adopted.

3. Promoting alternative water sources with a special emphasis on water harvesting and the use of non-conventional water sources.

4. Coordination between different intuitions and government bodies related to water sector.

5. Adopting systematic water quality investigation on the major water sources and combating water pollution, notably in surface water.

6. Applying economic water policies (e.g., tariffs, water metering, etc.) to mitigate the chaotic and irresponsible use of water.

7. Adaptation instruments must be implemented for the changing climatic conditions and their impact on water resources.

8. Applying efficient irrigation system to reduce the amount of water used for irrigation.

Author details

Amin Shaban
Lebanese National Council for Scientific Research, Beirut, Lebanon

*Address all correspondence to: geoamin@gmail.com

IntechOpen

© 2019 The Author(s). Licensee IntechOpen. This chapter is distributed under the terms of the Creative Commons Attribution License (http://creativecommons.org/licenses/by/3.0), which permits unrestricted use, distribution, and reproduction in any medium, provided the original work is properly cited.

References

[1] Shaban A, Hamze M. The Litani River, Lebanon: An Assessment and Current Challenges. Switzerland: Springer Science Publisher; 2018. 179 p. ISBN: 978-3-319-76300-2

[2] Darwic T, Shaban A, Hamze M. The national plan for Litani river remediation. In: The Litani River, Lebanon: An Assessment and Current Challenges. Switzerland: Springer Science Publisher; 2018. 179 p. ISBN: 978-3-319-76300-2

[3] Shaban A. Physical and anthropogenic challenges of water resources in Lebanon. Journal of Scientific Research and Reports. 2014, 2014;**3**(3):164-179

[4] CAL (Climatic Atlas of Lebanon). Meteorological Services. Vol. I. Lebanon: Ministry of Work and Public Transport; 1973. p. 31

[5] CAL (Climatic Atlas of Lebanon). Meteorological Services. Vol. II. Lebanon: Ministry of Work and Public Transport; 1982. p. 40

[6] TRMM (Tropical Rainfall Mapping Mission). Rainfall Archives. NASA. 2014. Available from: http://disc2.nascom.nasa.gov/Giovanni/tovas/TRMM_V6.3B42.2.shtml

[7] Lebanese Agricultural Research Institute (LARI). Climatic Data. Monthly Bulletin. Department of Irrigation and Agro-meteorology (DIAM); 2017, 16 p

[8] Funk C, Peterson P, Landsfeld M, Pedreros D, Verdin J, Shukla S, et al. The climate hazards infrared precipitation with stations—A new environmental record for monitoring extremes. Scientific Data. 2015;**2**:17. Article ID: 150066. DOI: 10.1038/sdata.2015.66

[9] Direction Générale de l'Aviation Civile (DGAC). Authority of Meteorology. Annual Report. Beirut, Lebanon; 1999. 32 p

[10] NOAA (National Oceanographic Data Center). Lebanon Climatological Data. Library. 2013. Available from: http://docs.lib.noaa.gov/rescue/data_rescue_lebanon.html

[11] El-Fadel M, Bou Zeid E. Climate Change and Water Resources in the Middle East: Vulnerability, Socio-economic Impacts, and Adaptation. The Fondazione Eni Enrico Mattei Note di Lavoro Series Index. 2001. Available from: http://www.feem.it/web/activ/_activ.html

[12] Somma J, Luxey P. Modélisation volumique des névés sur les hauts plateaux karstiques du Mont Liban. In: Actes de la conférence WATMED3 (Water Meditrranean) EU Project Publishing; Nov 1-3, 2006; Tripoli. 2006. 11 p

[13] Intergovernmental Panel for Climate Change, Working Group I (IPCC-WGI). Climate Change: The Science of Climate Change. Cambridge, UK: Cambridge University Press; 2010

[14] Shaban A. Support of space techniques for groundwater exploration in Lebanon. Journal of Water Resource and Protection. 2010;**5**:354-368

[15] Shaban A, Khawlie M, Abdallah C, Faour G. Geologic controls of submarine groundwater discharge: Application of remote sensing to north Lebanon. Environmental Geology. 2005;**47**(4):512-522

[16] Shaban A. Etude de l'hydroélogie au Liban Occidental: Utilisation de la télédétection [Ph.D. dissertation]. France: Bordeaux 1 Université; 2003. 202 p

[17] Shaban A, Hamze M. Shared Water Resources of Lebanon. New York: Nova

Science Publisher; 2017. 152 p. ISBN: 978-1-319-53612-111-7

[18] Regional Coordination on Improved Water Resources Management and Capacity Building. Regional Project. GEF, WB; National Council for Scientific Research, Lebanon (CNRS-L). 2015

[19] Index-Mundi. Lebanon Population Growth Rate. 2017. Available from: https://www.indexmundi.com/lebanon/population_growth_rate.html

[20] Shaban A. Analyzing climatic and hydrologic trends in Lebanon. Journal of Environmental Science and Engineering. 2011;**5**(3)

[21] Shaban A. New economic policies: Instruments for water management in Lebanon. Hydrology: Current Research. 2016;**7**(1):1-7

[22] Nehme N, Haidar C. Studying the physical and chemical and microbial characteristics of the Litani River water. In: The Litani River, Lebanon: An Assessment and Current Challenges. Switzerland: Springer Science Publisher; 2018. 179 p. ISBN: 978-3-319-76300-2

[23] Fadel A, Slim K. Evaluation of the physicochemical and environmental status of Qaraaoun reservoir. In: The Litani River, Lebanon: An Assessment and Current Challenges. Springer Science Publisher; 2018. 179 p. ISBN 978-3-319-76300-2

[24] Darwish T, Atallah T, Francis R, Saab C, Jomaa I, Shaaban A, et al. Observations on soil and groundwater contamination with nitrate, a case study from Lebanon-East Mediterranean. Agricultural Water Management. 2011. DOI: 10.1016/j.agwat.2011.07.016

Chapter 7

Process Modeling of Soil Thermal and Hydrological Dynamics

Nawa Raj Pradhan, Charles W. Downer and Sergei Marchinko

Abstract

To explicitly simulate the soil thermal state effects on hydrological response, the soil thermal regime, frozen soil, and permafrost simulation capability in the Geophysical Institute Permafrost Laboratory (GIPL) model have been included into the physically based, distributed watershed model Gridded Surface Subsurface Hydrologic Analysis (GSSHA). The GIPL model is used to compute a vertical soil temperature profile in every lateral two-dimensional GSSHA computational element using the soil moisture information from hydrologic simulations in GSSHA; GSSHA, in turn, uses this temperature and phase, ice content, and information to adjust hydraulic conductivities for both the vertical unsaturated soil flow and lateral saturated groundwater flow. This two-way coupling increases computational accuracy in both models by providing additional information and processes not previously included in either. The soil moisture physical state is defined by the Richards Equation, and the soil thermal state is defined by the numerical model of phase change based on quasi-linear heat conduction equation. Results from the demonstration site, a head water sub-catchment at the peak of the Caribou-Poker Creeks Research Watershed, representing Alaskan woodland and tundra ecosystem in permafrost-active region, indicated that freezing temperatures reduce soil thermal conductivity and soil storage capacity, thereby increasing overland flow and peak discharges.

Keywords: soil thermodynamics, soil hydrodynamics, soil temperature, soil moisture, soil hydraulic conductivity, soil infiltration

1. Introduction

Frozen soil has the potential to significantly affect hydrological response from hydrologic units of any scale, from the plot size to globally. In the United States, land surface freezing is an important hydrologic factor during winter in low temperature prevailing areas [1]. Approximately 60% of the Northern Hemisphere land surface is frozen in winter [2]. Frozen ground, heavy rains, and rapid snowmelt provide favorable conditions for extreme flooding [3]. Given the potential for significant consequences, frozen soil should be considered an important hydrologic factor in analysis of the potential for flooding, especially in the early spring when and where frozen ground, heavy rains, and rapid snowmelt prevail. Frozen soil hydrologic effects at larger scales, regional and global, include contribution of net freshwater to the Arctic Ocean [4], affecting ocean salinity and global thermohaline circulation [5]. While global and regional hydrologic and climate studies require a realistic representation of the hydraulic and thermal properties [6], a physical process-based representation of these properties starts from a small, point, or field scale. Plot-scale studies by Dunne and Black [7] and Stähli et al. [8], among others, show that a catchment's rainfall runoff generation process is altered when infiltration is reduced during the transition of the soil water phase toward the freezing state. As the distribution of frozen soils is highly variable in both space and time [9], to fully understand both the occurrence of frozen soil and the subsequent impact on hydrology requires a physics-based, distributed analytical tool that provides a two-way coupling of the soil thermodynamic and hydrologic state to provide insight into the temporal and spatial variability of soil thermal state and its effects on the time and space distributions of the hydrological response.

In this chapter we describe the development of a coupled soil thermodynamic/hydrologic model that accounts for the physical processes of thermodynamics, the interaction of the thermodynamics with hydrodynamics, and the variability of freezing condition and runoff in space and time utilizing two well-known and widely applied physics-based models for soil thermodynamics and watershed hydrology, GIPL [10, 11], and GSSHA [12, 13]. This coupled model is demonstrated on a contrived watershed, around a previous GIPL test site [13]. In this demonstration study, the coupled model was deployed in the headwater sub-catchment of the Caribou-Poker Creeks Research Watershed.

2. Development of thermo-hydrodynamic model

The coupled model was developed from two widely known, documented, and applied based models, with applications for permafrost and watershed analysis, the Geophysical Institute Permafrost Laboratory (GIPL) model and the Gridded Surface Subsurface Hydrologic Analysis (GSSHA) model. A brief description of the models and the relevant processes are described in the following sections.

2.1 Watershed hydrological model

The GSSHA model is the basis for the overall model development. The GSSHA model was chosen because it is a published, fully distributed, physics-based, watershed model with wide applicability for computing watershed system states, such as soil moisture, snow cover, overland flow depth, stream discharge, and groundwater heads. The GSSHA works on a uniform grid. Any number of point processes, such as distributed precipitation, snow accumulation and melting, evapotranspiration, infiltration, groundwater recharge, etc., can be computed [14–18]. Point processes can be integrated to produce the system response, such as discharge, groundwater flow, etc. GSSHA is an option-driven model, and many processes can be simulated with multiple methods. The focus here is on the subsurface processes.

GSSHA simulates both saturated and unsaturated water movements. In reality, subsurface flow is three-dimensional with complex layering, varying free surface, confined aquifers, and other complications. In GSSHA, the subsurface is represented in a physically explicit manner consistent with the main purpose of GSSHA, which is to simulate surface water flows and the interaction of the surface and subsurface systems. Saturated and unsaturated flows are solved in separate domains. The two domains are linked with head and flux boundaries.

2.1.1 Unsaturated zone model

The unsaturated, or vadose, zone controls the flux of water from the land surface to the saturated groundwater zone and determines the partitioning of water into runoff, infiltration, ET, and groundwater recharge. GSSHA provides many ways to analyze these processes including an integrated solution of soil water movement and state described by the Richards Equation [19]. The Richards Equation [19] is considered the most physically correct mathematical representation of the vadose zone. While flow in the vadose zone is in three dimensions, the predominant direction of flow is vertical. In GSSHA, the 1D, vertical, head-based form of the Richards Equation is solved [20, 21]:

$$C_m(\psi)\frac{\partial \psi}{\partial \tau} - \frac{\partial}{\partial z}\left[K_{soil}(\psi)\left(\frac{\partial \psi}{\partial z} - 1\right)\right] - W = 0 \tag{1}$$

where C_m is the specific moisture capacity, ψ is the soil capillary head (cm), z is the vertical coordinate (cm) (downward positive), τ is time (h), $K_{soil}(\psi)$ (cm) is the effective hydraulic conductivity, and W is a flux term added for sources and sinks (cm h^{-1}), such as ET and infiltration. The head-based form is valid in both saturated and unsaturated conditions [22].

Heads for each cell as first computed using an implicit central difference in space and forward difference solution and then flux updating are performed to ensure mass balance. The variables K_{soil} and C_m in Eq. (1) are nonlinear and depend on the water content of each cell. K_{soil} and C_m are calculated employing Brooks and Corey [23] equations, as extended by Huston and Cass [24].

2.1.2 Saturated groundwater model

While groundwater flow can be free surface or confined, and may be three-dimensional, GSSHA solves a 2D lateral solution of the free surface groundwater flow equations as described by Pinder and Bredehoeft [25]:

$$\frac{\partial}{\partial x}\left(T_{xx}\frac{\partial h}{\partial x}\right) + \frac{\partial}{\partial x}\left(T_{xy}\frac{\partial h}{\partial y}\right) + \frac{\partial}{\partial y}\left(T_{yx}\frac{\partial h}{\partial x}\right) + \frac{\partial}{\partial y}\left(T_{yy}\frac{\partial h}{\partial y}\right) = S\frac{\partial h}{\partial \tau} + W(x, y, \tau) \tag{2}$$

where T is the transmissivity ($m^2\ s^{-1}$), h is the hydraulic head (m), S is the storage term (dimensionless), and W is the flux term for sources and sinks ($m\ s^{-1}$).

Downer [20] simplifies Eq. (2) by assuming no directional heterogeneity in the transmissivity terms, expressing transmissivity as the product of the hydraulic conductivity of the media (K_{soil}) and the depth of the saturated media (b). Substituting surface water elevation (E_{ws} = h + datum) for the head, the free surface problem can be described as [20]

$$\frac{\partial}{\partial x}\left(K_{xx}b\frac{\partial E_{ws}}{\partial x}\right) + \frac{\partial}{\partial y}\left(K_{yy}b\frac{\partial E_{ws}}{\partial y}\right) = S\frac{\partial E_{ws}}{\partial \tau} + W(x, y, \tau) \tag{3}$$

Details of the solution can be found in Downer [20].

2.2 Soil thermodynamic model

GIPL was developed at the Geophysical Institute Permafrost Laboratory for the explicit purpose of determining soil temperature profiles and locating the areas of frozen soil in the profile, including the permafrost. GIPL is a stand-alone 1D soil thermal and permafrost model that is used to compute a one-dimensional (vertical) soil temperature profile over time using static values of soil moisture.

The Stefan problem [26, 27] with phase change is the problem of thawing or freezing via conduction of heat. GIPL solves the Stefan problem employing the enthalpy formulation. One-dimensional, vertical, quasi-linear heat conduction Equation [28] is the basis of the GIPL numerical model:

$$\frac{\partial H(x,t)}{\partial \tau} = \nabla \cdot (k(x,t)\nabla t(x,\tau)) \tag{4}$$

where x is a vertical spatial variable which ranges between x_u, upper depth of the computational unit, and x_L, lower depth of the computational unit. τ is the time and t is the temperature. $k(x,t)$ is a thermal conductivity (W m^{-1} K^{-1}). $H(x,t)$ is an enthalpy function defined as

$$H(x,t) = \int_0^t C(x,s)ds + L\theta(x,t) \tag{5}$$

where L is the volumetric latent heat of freeze/thaw (MJ m^{-3}), $C(x,s)$ is the volumetric heat capacity (MJ m^{-3} K^{-1}), and $\theta(x,t)$ is the volumetric unfrozen water content (fraction of the total water content). Boundary and initial conditions are required in solving Eq. (4). The boundary condition on the upper extent of the domain corresponds to the near land surface air layer. Embedding seasonal snow layer into the air layer is allowed by the fictitious domain formulation [29]. The upper boundary condition is defined as the Dirichlet-type boundary condition:

$$t(x_u, \tau) = t_{air} \tag{6}$$

where t_{air} is a daily averaged air temperature. The lower boundary is set as the geothermal gradient as

$$\frac{\partial t(x_l, \tau)}{\partial x} = g \tag{7}$$

where g is the geothermal gradient, a small constant (km^{-1}). For the initial temperature distribution, an appropriate ground temperature profile based on the point location is used:

$$t(x,\tau) = t_0(x) \tag{8}$$

The unfrozen water content $\theta(x,t)$ is defined as:

$$\theta(x,t) = \eta(x)\begin{cases} 1, & t \geq t_* \\ a|t_* - t|^{-b}, & t < t_* \end{cases} \tag{9}$$

$\eta(x)$ is a volumetric soil moisture content. a and b are dimensionless positive constants [30]. The constant t_* is a freezing point depression, the temperature at which ice begins to form in the soil. The depth and time variation of the unfrozen water content $\theta(x,t)$ depends on hydrologic forcing and soil type. The numerical solution in GIPL is an implicit, finite difference scheme. A detailed mathematical description of the model and numerical solution methods of Eq. (4) can be found in Marchenko et al. [11], Sergueev et al. [28], and Nicolsky et al. [31]. GIPL input data include soil thermal properties, lithological data and vegetative cover, climate data, and snow cover.

2.3 Coupling GIPL to GSSHA

The linkage of the soil thermal and water movement solutions in GSSHA facilitates the temperature domain solution based on the varying soil moisture and the soil water movement domain solutions adjusted for the varying soil temperature condition. In linking the GIPL model to GSSHA, GIPL is essentially included in GSSHA as another point process. The GIPL model is used to compute a vertical soil temperature profile in every lateral two-dimensional GSSHA computational element using the soil moisture information from hydrologic simulations in GSSHA; GSSHA, in turn, uses this temperature and water phase information to adjust hydraulic conductivities for both the vertical unsaturated soil flow and lateral saturated groundwater flow. This two-way coupling increases computational accuracy in both models by providing additional information and processes not previously included in either.

As shown in **Figure 1**, the GSSHA model provides the spatial variability of land surface and hydrodynamic parameters such as air temperature, soil properties, subsurface soil moisture state, etc. to GIPL. GIPL employs the information provided by GSSHA to update the soil thermal state and pass it back to GSSHA. This thermal state information is employed to determine whether the soils are frozen or not. This frozen or unfrozen information is employed by GSSHA to adjust hydraulic conductivities, hydraulic transmissivity, and soil moisture state and saturation levels used in water flow computations. These updated soil saturation and soil moisture state are then employed by the GIPL model to produce new thermal state profiles. This exchange of information continues throughout simulation duration, as shown in **Figure 2**.

The implementation of GIPL into GSSHA provides a robust soil temperature/moisture tool with many improvements compared to the stand-alone version of GIPL. In GSSHA, the 1D GIPL model is used to compute a soil temperature profile in every 2D GSSHA grid cell, providing a 3D map of soil temperature/water state. As implemented in GSSHA, GIPL no longer utilizes a static time step nor soil water

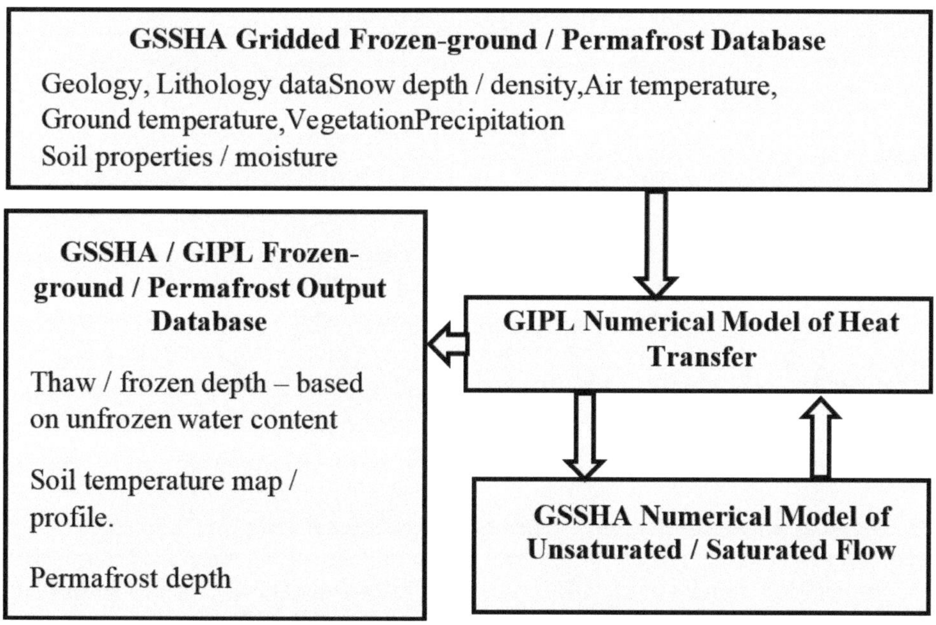

Figure 1.
GIPL as a frozen ground/permafrost component in GSSHA.

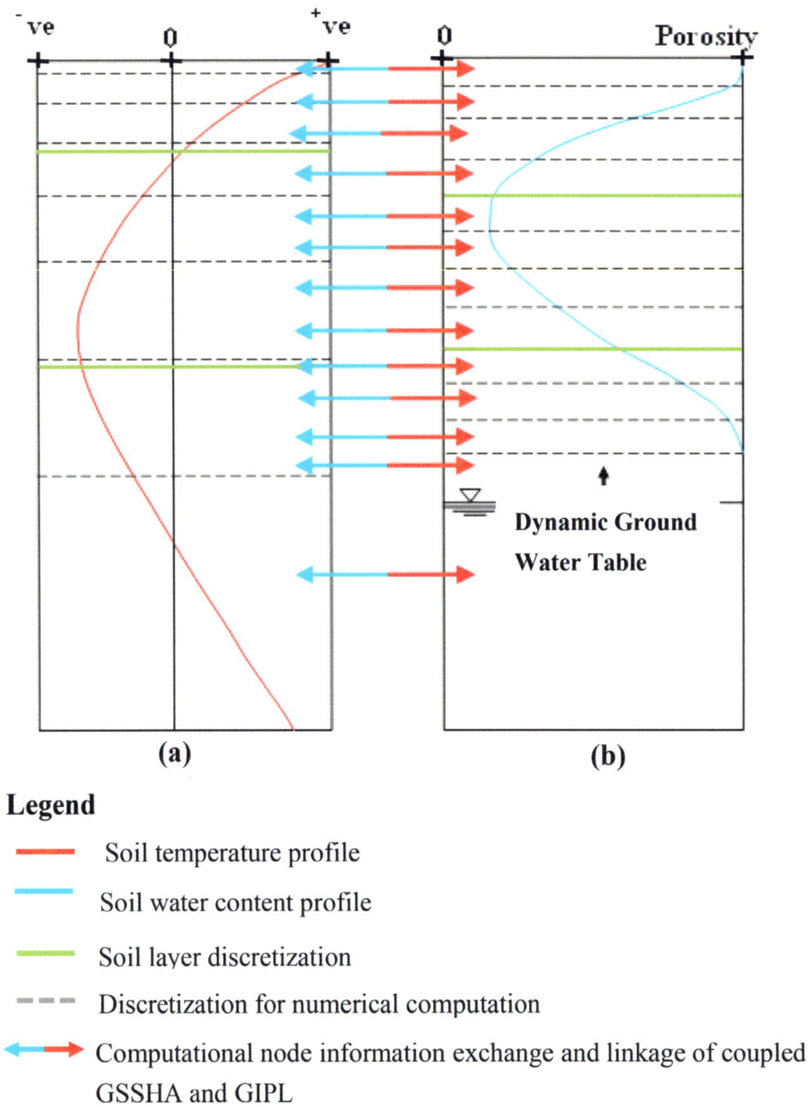

Figure 2.
GSSHA and GIPL coupling architecture.

state. Instead, GIPL performs soil temperature computations utilizing the time and space varying soil moistures computed using the Richards equation. The time step of GIPL is set to the overall GSSHA model time step, which is on the order of minutes, as opposed to days. The GSSHA input formats are used to specify 3D distributed soil parameter values for the GIPL solution. Several thermo-hydrodynamic formulations and modeling concepts are implemented to link and exchange the information in GIPL and GSSHA.

2.3.1 Linking soil temperature and soil water computational nodes

The solution domain of the GIPL soil temperature model overlaps in a somewhat complex manner with both the saturated and unsaturated soil water movement domains in GSSHA. In no-flux lower GIPL boundary condition, the GIPL domain must extend very deep into the soil/permafrost, as much as 1000 m or more.

In GSSHA, only the surficial aquifer is simulated, so the saturated groundwater domain is down to the first confining layer in the subsurface. This is typically on the order of a few to hundreds of meters deep. The unsaturated zone domain is any soil above the saturated zone. The unsaturated domain is dynamic in both space and time and can vary from no domain (groundwater table is at or above the soil surface) to the depth of the surficial aquifer, depending on groundwater conditions. The unsaturated zone is further divided into four regions, corresponding to the A, B, and C soil horizons, as well as the deeper groundwater media.

Because of the differences in domains, and requirements for solution, in the coupled framework, the GIPL domain and discretization are independent of the saturated and unsaturated soil water domains and discretizations. The linkage of computational nodal discretized information from GIPL to GSSHA and vice-versa is shown in **Figure 2**.

2.3.2 Soil temperature effect on hydraulic conductivity

The primary effect of soil temperature on the soil water domain is that freezing soils result in much lower hydraulic conductivities of the soil. In the unsaturated zone, the vertical soil hydraulic conductivity is computed from the relative saturation (S_E), a nondimensional parameter that varies between 0 and 1.

The relative fraction of liquid water of the total soil moisture, S_E, can be computed from the soil temperature as [32]

$$S_E = \left(\frac{1}{1 + (\propto|1.22.t|)^n}\right)^m \textit{for } T \leq 0^\circ C \tag{10}$$

where n, m, and α are the van Genuchten parameters; t is the soil temperature in °C. S_E is always 1 for temperatures above 0°C. For temperatures below −10°C, the value of S_E is assumed to be 0.

As soil freezes, pathways through the soil, pores, close reducing the ability of the soil to transmit water. This results in a reduction of the soil hydraulic conductivity. In the unfrozen portion of the soil, an exponential response in effective hydraulic conductivity has been measured for freezing/thawing mineral and organic soils [33]. The temperature-adjusted relative saturation, S_E, can be used to compute the soil temperature-adjusted hydraulic conductivity as the sum of the hydraulic conductivity of the unfrozen pores and the hydraulic conductivity of the frozen pores:

$$K_{soil}(t) = e^{S_E ln(K_t) + (1 - S_E)K_f} \tag{11}$$

where $K_{soil}(t)$ is the effective hydraulic conductivity in m s^{-1}, K_t is the hydraulic conductivity for S_E = 1, and K_f is the frozen hydraulic conductivity (S_E = 0). The frozen soil has a very little capacity to transmit water. In GSSHA K_f is set to 10^{-6}.

2.3.3 Soil heat transfer effect on effective groundwater transmissivity

The effect of soil temperature on the saturated groundwater solution is considered by adjusting the transmissivity of the groundwater based on the frozen water content in each groundwater cell. Because of soil structure, lateral hydraulic conductivities are essentially 10 to 100 times larger than vertical hydraulic conductivities. In practical terms, frozen saturated media has little to no ability to laterally transport water compared to the unfrozen media.

In the soil thermal calculations, the full profile of soil temperatures allows the determination of discrete soil thermal cells that are below freezing and thus incapable of transmitting saturated soil water laterally. However, in the 2D saturated flow calculations, there is no vertical discretization of the domain. To account for the effect of the frozen sections in the groundwater media, the depth of flow (b) [Eq. (3)] is adjusted by subtracting the total frozen portions of the unsaturated zone, contiguous or not, to compute an effective flow depth ($b_{effective}$). The effect of this reduced the transmissivity in Eq. (2), which is the primary control on flow in the equation.

The depth of the unfrozen saturated media in GSSHA is identified by determining which soil thermal computational nodes correspond to the saturated media depth and then adding up all unfrozen, $T > 0$, cells to determine the unfrozen saturated flow depth ($b_{effective}$). If the groundwater surface is between frozen and unfrozen thermal cells, then the portion unfrozen is determined by interpolation. This avoids the overestimation of the effective saturated depth.

Once the effective saturated depth is calculated, local−/grid-based groundwater transmissivity is defined as.

$$T = K_{soil}{}^{*}\ b_{effective} \tag{12}$$

3. Study area and data

3.1 Description of the study site

Figure 3(c) shows the location of the study area with elevation and gridded model domain. The study area is in the Caribou-Poker Creeks Research Watershed (CPCRW), 48 km north of Fairbanks with latitude of 65°10′ N and longitude of 147° 30′ W Alaska. The CPCRW, which encompasses an area of 101.5 km^2 within the boreal forest, is part of the Long-Term Ecological Research (LTER) network. Parts of this watershed are underlain by discontinuous permafrost [34].

There is a weather station on the summit of Caribou Peak as shown in **Figure 3** which is called CPEAK and is at an elevation of 773 m. This station has a 10-meter tower with various atmospheric sensors and ground temperature thermistors at several depths (http://www.lter.uaf.edu/data/metadata/id/442/inline/1).

As shown in **Figure 3**, A 10 by 10 GSSHA/GIPL gridded model was developed for simulations in the study area. This study area model domain and location is selected as to include the CPEAK station so that the observations from the station could be used in model development and validation.

3.2 Data types and sources

The Caribou-Poker Creeks Research Watershed (CPCRW) is one of the study sites for the Long-Term Ecological Research (LTER) project where long-term observed data are maintained by the Institute of Arctic Biology at the University of Alaska, Fairbanks, and made available online (http://www.lter.uaf.edu/data). The website has a search filter page where the hydrology and climate data for the study sites is available. The data from CPEAK was employed as initial condition of soil profile temperature and soil moisture and hydrometeorological forcings/input to run the model.

3.2.1 Topography, soil, and land use

A 50-meter digital elevation model (DEM), obtained from the National Elevation Dataset (NED) and downloaded through the National Map Viewer http://na

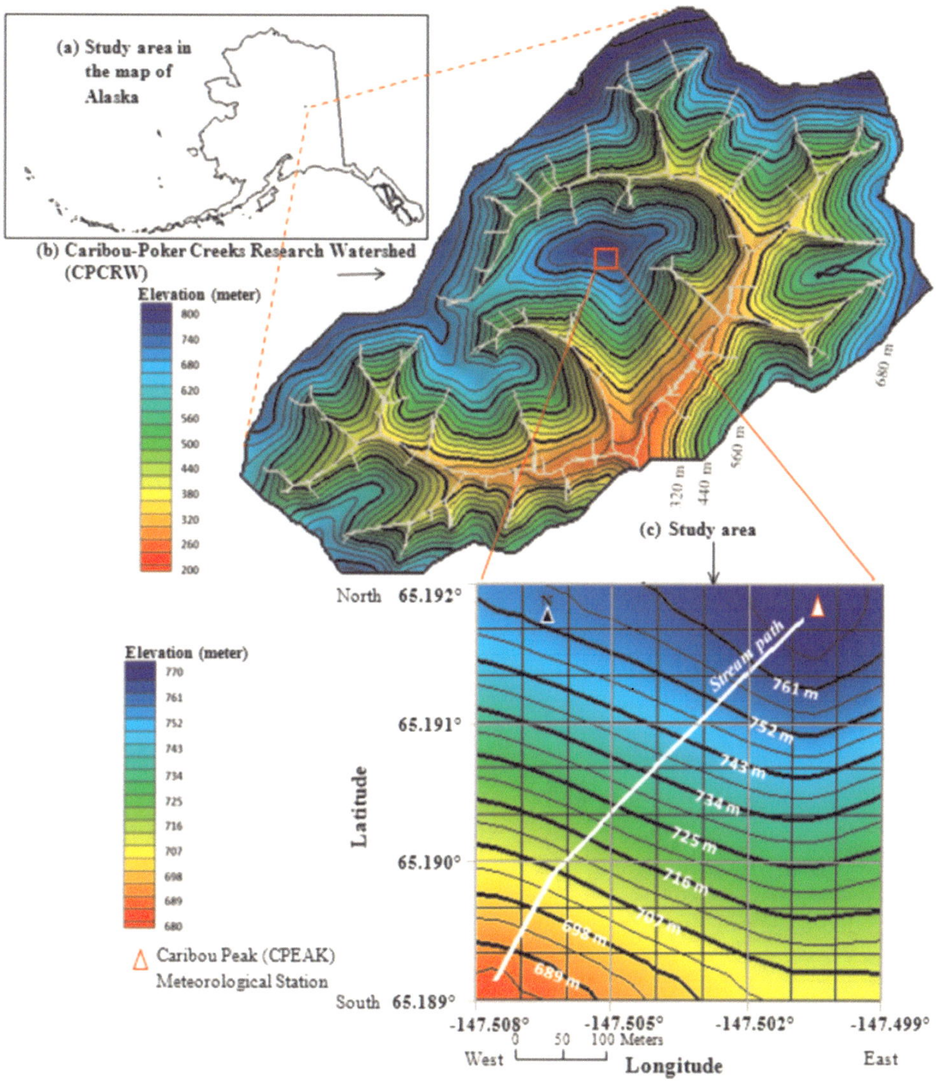

Figure 3.
Study area.

tionalmap.gov/viewer.html, was employed to develop the study area model shown in **Figure 3**.

Figure 4(a) shows the soils of the study area, and the soil description, as per Rieger et al. [35], is shown in **Table 1**. The study area vegetation-type distribution, based on a unified statewide system for classifying vegetation in Alaska [36], is shown in **Figure 4(b)**.

3.2.2 Climate of the study area

CPCRW lies in the interior climatic division of Alaska characterized by low annual precipitation, low cloudiness, low humidity, and large diurnal and annual temperature variations. In the study region, the 30-year normals [37] show that January is typically the coldest month, with a mean temperature of −24.4°C, and July, the warmest month, with a mean temperature of 17.1°C. The average annual precipitation in this region is 285 mm, and the wettest months are June, July, and

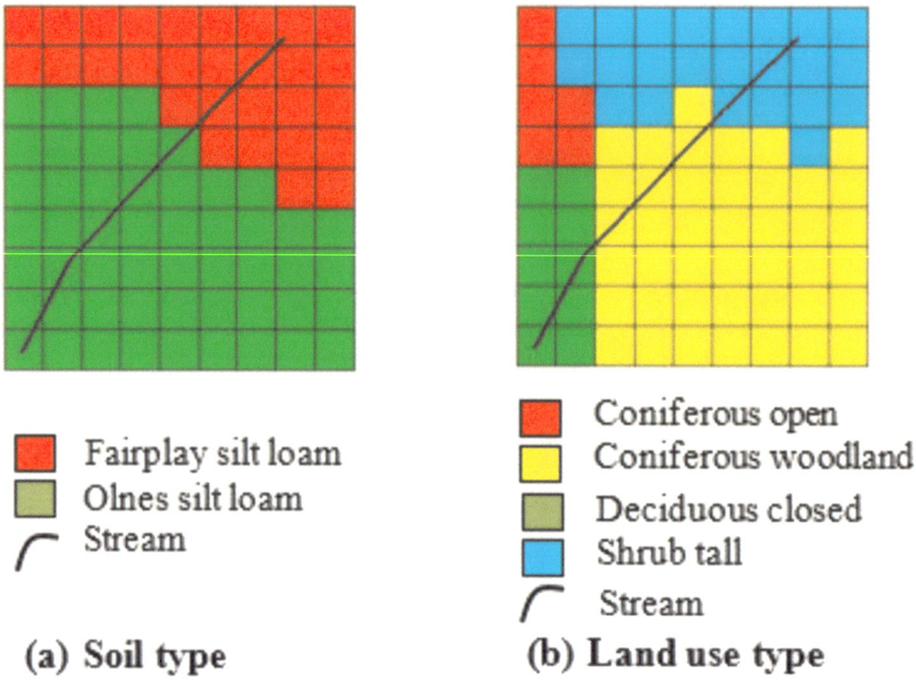

Figure 4.
Soil types and vegetation types of the study area.

Soil series	USDA texture	Drainage
Fairplay	Silt loam and gravelly silt loam	Moderately well drained
Olnes	Silt loam and very gravelly silt loam	Well drained

Table 1.
Soil properties of the study area.

August [38]. Annual snowfall averages 1692 mm and commonly covers the ground from October to April [37]. The model employed CPEAK hydrometeorological data of the year 2002 included relative humidity (%), wind speed (kts), air temperature (°F), barometric pressure (in Hg), solar radiation (W h m^{-2}), and tipping bucket rainfall rates and are available at http://www.lter.uaf.edu/data.

4. Model development of the study area

A 10 by 10 gridded model of the study area, shown in **Figure 3**, includes precipitation, overland flow, unsaturated zone computations using the 1D Richards equation, ET using the Penman-Monteith Equation [39, 40], and 1D soil thermal computations using GIPL.

4.1 Initial soil moisture and soil temperature conditions

The initial temperature condition which is from CPEAK ground temperature thermistors at several depths on 2002-5-1, 1 AM, is shown in **Figure 5**. The numerical solution of the soil thermal state using quasi-linear heat conduction equation (Eq. (4)) employs this initial temperature condition of the soil profile.

The initial soil moisture condition obtained from Caribou-Poker Creeks Research Watershed for 2002-5-1, 1 AM [41], is shown in **Figure 6**. This initial soil moisture condition of the soil profile is employed by the numerical solution of the soil moisture, Richards equation, in the unsaturated vadose zone (Eq. (1)).

4.2 Model parameter values

The model parameter values, distributed on grids horizontally and on soil layers vertically, for processes, such as overland flow, infiltration, evapotranspiration, and

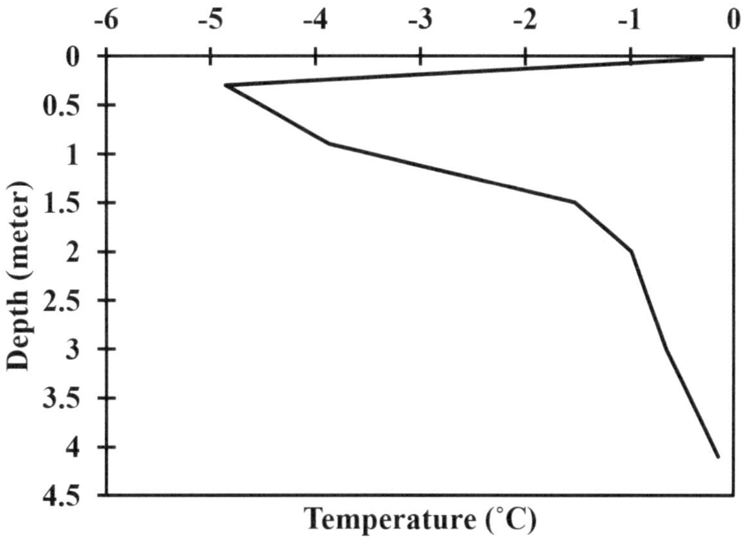

Figure 5.
Initial soil temperature condition.

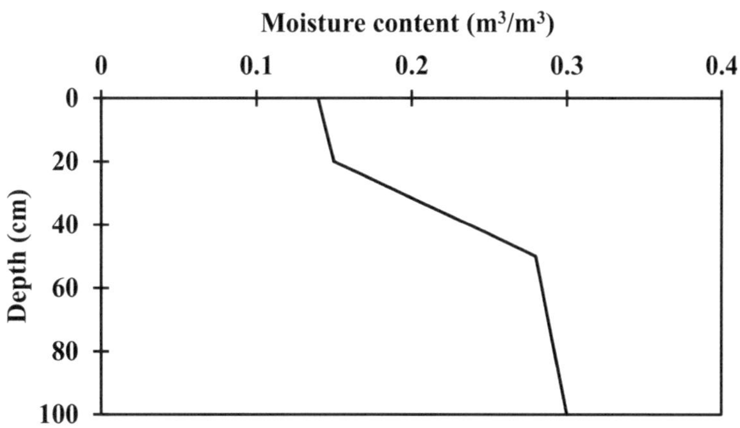

Figure 6.
Initial soil moisture condition.

Soil heat conductivity (W/m K)	Soil volumetric heat capacity (J/m^3 K)
3.0	2,800,000

Table 2.
Soil thermal parameter values.

thermodynamics, are assigned based on the soil texture, shown in **Figure 4a**, and land use, shown in **Figure 4b**.

Table 2 shows the soil thermal parameters with values, representing the silt loam [31, 42, 43] of Alaskan woodland and tundra ecosystem sites in permafrost-active region, employed in the thermodynamics process.

Table 3 shows Manning's roughness parameter values, representing the vegetation type of **Figure 4(b)** [44, 45], employed in the routing process.

Table 4 shows the soil hydraulic parameters with values, representing the silt loam ([46], https://www.gsshawiki.com) of **Figure 4(a)**, employed in the infiltration and soil water retention process.

In this study, Penman-Monteith method was employed as evapotranspiration process. **Table 5** shows the Penman-Monteith parameters that includes vegetation transmission coefficient (light penetration through canopy), values of land surface albedo, vegetation height (for aerodynamic resistance term), and vegetation canopy resistance (for stomatal control of the loss of water). **Table 5** shows the ET

Land use type	Manning's roughness values
Coniferous open	0.17
Coniferous woodland	0.19
Deciduous closed	0.2
Shrub tall	0.25

Table 3.
Manning's roughness values.

Infiltration parameter	Soil layer		
	Top	Middle	Lower
Soil thickness (cm)	20	50	—
Saturated hydraulic conductivity (cm/h)	0.5	0.5	0.5
Pore size distribution index	0.60	0.694	0.694
Wetting front suction head (cm)	8.0	6.0	6.0
Porosity (m^3/m^3)	0.42	0.40	0.40
Residual soil moisture content (m^3/m^3)	0.04	0.045	0.045

Table 4.
Soil parameter values for the Richards infiltration scheme.

Land use type	Albedo	Vegetation height (cm)	Canopy resistance (s/m)	Transmission coefficient
Coniferous open	0.15	10	120	0.18
Coniferous woodland	0.15	10	120	0.18
Deciduous closed	0.2	12	200	0.18
Shrub tall	0.2	1.3	150	0.5

Table 5.
Evapotranspiration parameter values.

parameter values, representing the vegetation type of **Figure 4(b)**, employed in the evapotranspiration process.

All the parameter values defined in **Table 5** were taken from the literature, which are also defined in the GSSHA wiki https://www.gsshawiki.com. The literature values for albedo and vegetation height are defined in Eagleson [47]. The literature values for canopy resistance are defined in Monteith [39]. The literature values for the transmission coefficient are defined in Sutton [48].

5. Result and discussion

5.1 Soil thermodynamics

The simulation period was from May 1st to May 31st of 2002, a period during which air temperatures are beginning to rise above freezing. The daily maximum soil temperature obtained from the simulation is compared with the observed one in **Figure 7**. Both the observed and simulated temperatures in **Figure 7** is effective at a depth of 10 cm in the soil profile. The root-mean-square error for this daily maximum soil temperature was 4.7°C. It was found that the soil thermal

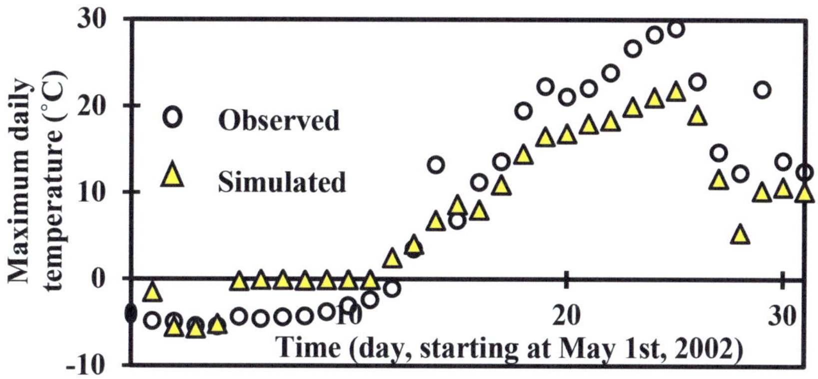

Figure 7.
Comparison of the time series of observed and simulated temperature.

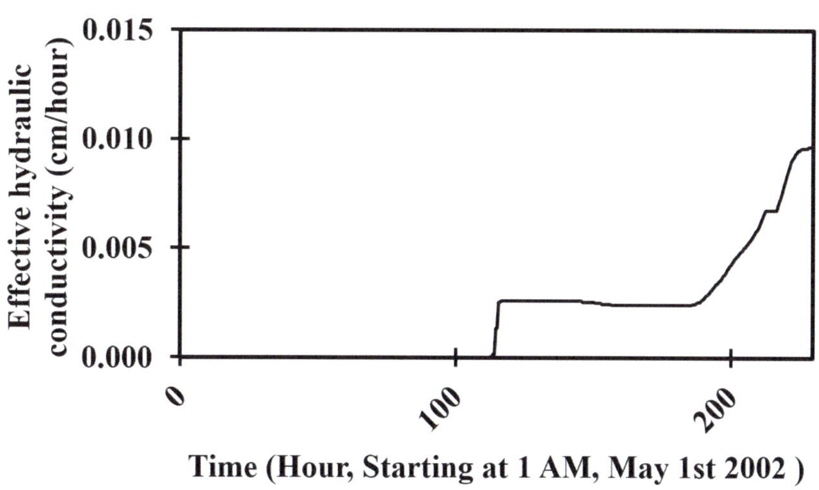

Figure 8.
Hydraulic conductivity under freezing and thawing soil active layer.

conductivity parameter value was most sensitive and effective for the near-surface simulated temperature. It was also found that the temperature of the near-surface soil layer is directly influenced by the air temperature.

5.2 Soil's effective hydraulic conductivity

Figure 8 shows the simulation evolution of the effective hydraulic conductivity starting on May 1st of 2002. There are no observed hydraulic conductivity values to compare to, but the model intuitively represents the condition. Initially, the hydraulic conductivity in **Figure 8** is very low, almost zero in the first 113 hours, the result of reduced effective saturation in the soil due to the soil being completely frozen (**Figure 7**). As the air and soil temperatures rise, the soil begins to thaw, with a resulting increase in hydraulic conductivity. The exponential rise of hydraulic conductivity is a consistent observation from freezing/thawing mineral and organic soils [33]. The simulated effective hydraulic conductivity in **Figure 8** is from Eq. (11), where the effective soil hydraulic conductivity K_{soil} at a given temperature (t) is a function of hydraulic conductivity of the unfrozen soil and the effective saturation S_E.

5.3 Hydrologic runoff response

Figure 9 shows the comparison of GSSHA simulated discharge with and without taking into account freezing and thawing soil properties in the study area. While there are no measurements of runoff from the site, the results are consistent with the results presented for air and soil temperature and resulting hydraulic conductivity shown above. The freezing soil temperature (**Figure 7**) leads to increased coverage of frozen soil which in turn leads to less soil pore water storage. The reduced soil pore water storage capacity leads to decrease hydraulic conductivity as

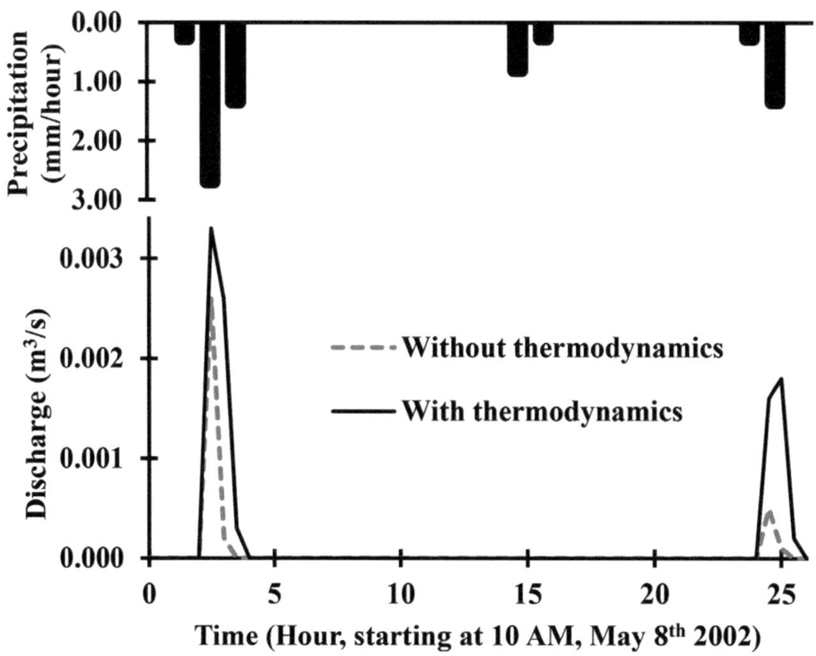

Figure 9.
Hydrograph with and without thermodynamics.

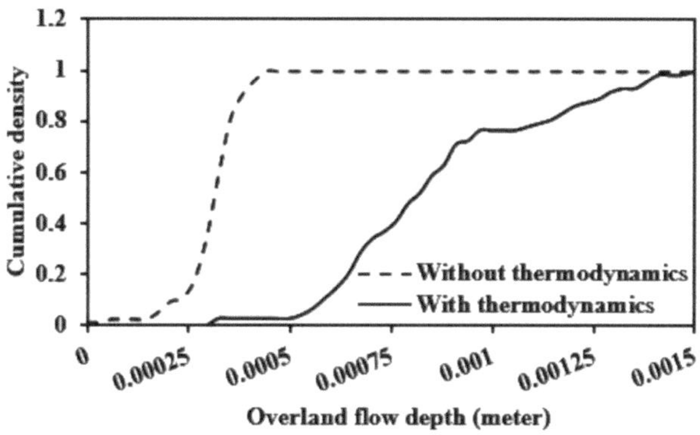

Figure 10.
Comparison of distribution of overland runoff with and with thermodynamic process in the model simulation.

shown in **Figure 8**, which results in a flashier response to the precipitation event as shown in **Figure 9** graph with thermodynamics. On the other hand, loss of frozen soil and permafrost or without taking into account thermodynamics will lead to enhanced connectivity between the surface and groundwater storage regimes and decreased overland flow as shown by the graph without thermodynamics in **Figure 9**.

5.4 Discussion

Unless the simulation discharge represents the distributed physical runoff process in a realistic way, even a well-calibrated simulated discharge at a catchment outlet may only be a right answer for wrong reasons. The modeling and the simulations in this study have explicitly taken into account frozen soil as an important hydrologic factor. The simulation results showed the variability of freezing condition in space and time. How this variability of freezing condition in space and time would affect the distribution of overland runoff is particularly important from the concern of climate change and land use change effect in local hydrology [7, 49]. **Figure 10** shows the model simulated distribution of spatially distributed overland runoff, at the peak of the precipitation event of May 7th and May 8th of 2002 in CPEAK station, by taking into account thermodynamics and without taking into account thermodynamics. It is clear from **Figure 10** that the simulation significantly underestimated the runoff when thermodynamic process was absent.

6. Recommendations and conclusions

A coupled framework for simulating the interaction between soil temperature, including permafrost, and hydrology was developed by incorporating the soil temperature and permafrost model GIPL into the distributed, physics-based hydrologic model GSSHA. This chapter describes the architecture for numerically linking the GIPL thermodynamic model into GSSHA's hydrodynamic modeling framework. Deploying this enhanced capability showed that GSSHA hydrodynamics include soil moisture saturation feedback in the vadose zone as well as the corresponding soil ice content effects on hydraulic conductivity and transmissivity. In this study the coupled model was deployed in the headwater region at the peak of the Caribou-Poker Creeks Research Watershed that included the hydrometeorological and soil property measurement station called CPEAK.

The model captured the seasonal rise of soil temperatures and thaw of frozen soils. The model showed intuitively correct representations of soil hydraulic conductivity and runoff, consistent with the observed rise of soil temperatures. Numerical simulations showed the hydrologic importance of frozen soils with the implication that climate change could have large effects on hydrology as air and soil temperatures rise near the poles resulting in loss of permafrost and increase seasonal thawing of soils.

The main use of the "process modeling of soil thermal and hydrological dynamics" has been to generate spatial and temporal dataset of permafrost distribution and ground temperature dynamics as well as the active layer thickness, the depth of the seasonal thaw. Such dataset would be useful in the assessments of a wide range of thermo-hydrodynamic related fields, including ecology, climatology, and socio-economy, in the cold regions.

Acknowledgements

Pradhan et al. [13] was the initial hydrodynamic and thermodynamic model integration effort which was then funded by Strategic Environmental Research and Development Program (SERDP) under Project Number 11 RC-2110. Theoretical and technical contributions of Dr. Anna Liljedahl, at the University of Alaska Fairbanks are significant for that initial hydrodynamic and thermodynamic model integration effort. Article Processing Charges (APCs) for the publication were supported by the U.S. Army Corps of Engineers Flood and Coastal Storm Damage Reduction Research and Development Program under the technical direction of Dr. Julie Rosati and Mr. Ian E. Floyd. The Bonanza Creek Long-Term Ecological Research program (BNZ LTER) at the University of Alaska Fairbanks is thanked for making and sharing the hydrology and climate data set of the CPCRW in Alaska. Constructive comments from anonymous reviewers are greatly appreciated.

Author details

Nawa Raj Pradhan[1]*, Charles W. Downer[1] and Sergei Marchinko[2]

1 Coastal and Hydraulics Laboratory, U.S. Army Engineer Research and Development Center, Vicksburg, MS, USA

2 Geophysical Institute, University of Alaska Fairbanks, USA

*Address all correspondence to: nawa.pradhan@usace.army.mil

IntechOpen

© 2019 The Author(s). Licensee IntechOpen. This chapter is distributed under the terms of the Creative Commons Attribution License (http://creativecommons.org/licenses/by/3.0), which permits unrestricted use, distribution, and reproduction in any medium, provided the original work is properly cited. CC BY

References

[1] Storey HC. Frozen soil and spring and winter floods. In: Stefferud A, editor. Water: The Yearbook of Agriculture 1955. Washington, D.C: U.S. Department of Agriculture; 1955. pp. 179-184

[2] Zhang T, Barry RG, Knowles K, Heginbottom JA, Brown J. Statistics and characteristics of permafrost and ground ice distribution in the northern hemisphere. Polar Geography. 1999; **23**(2):147-169

[3] Hirschboeck KK. Climate and floods. In: National Water Summary, 1988–89 —Floods and Droughts: Hydrologic Perspectives on Water Issues. USGS Water-Supply Paper 2375. 1991. p. 591

[4] Barry RG, Serreze MC. Atmospheric components of the Arctic Ocean freshwater balance and their interannual variability. In: Lewis EL et al., editors. The Freshwater Budget of the Arctic Ocean. Dordrecht, Netherlands: Springer; 2000. pp. 45-56

[5] Broecker WS. Thermohaline circulation, the Achilles heel of our climate system: Will man-made CO2 upset the current balance? Science. 1997; **278**:1582-1588

[6] Niu G, Yang Z. Effects of frozen soil on snowmelt runoff and soil water storage at a continental scale. *Journal of Hydrometeorology*. 2006;7:937-952. DOI: 10.1175/JHM538.1

[7] Dunne T, Black RD. Runoff processes during snowmelt. Water Resources Research. 1971;7:1160-1172

[8] Stähli M, Jansson PE, Lundin LC. Soil moisture redistribution and infiltration in frozen sandy soils. Water Resources Research. 1999;**35**:95-103

[9] Stähli M. Hydrological significance of soil frost for pre-alpine areas. Journal of Hydrology. 2017;**546**:90-102

[10] Jafarov EE, Marchenko SS, Romanovsky VE. Numerical modeling of permafrost dynamics in Alaska using a high spatial resolution dataset. The Cryosphere Discussions. 2012;**6**:89-124. DOI: 10.5194/tcd-6-89-2012. Available from: www.the-cryosphere-discuss.net/6/89/2012/

[11] Marchenko S, Romanovsky V, Tipenko G. Numerical modeling of spatial permafrost dynamics in Alaska. In: Proceedings of the Eighth International Conference on Permafrost; 190–204; Willey, Institute of Northern Engineering. Fairbanks: University of Alaska; 2008. pp. 91-95

[12] Downer CW, Ogden FL. Gridded Surface Subsurface Hydrologic Analysis (GSSHA) User's Manual; Version 1.43 for Watershed Modeling System 6.1, System Wide Water Resources Program, Coastal and Hydraulics Laboratory, U.S. Army Corps of Engineers, Engineer Research and Development Center, ERDC/CHL SR-06-1; 2006. 207 p

[13] Pradhan NR, Downer CW, Marchenko S, Lijedahl A, Douglas TA, Byrd A. Development of a Coupled Framework for Simulating Interactive Effects of Frozen Soil Hydrological Dynamics in Permafrost Regions, ERDC TR-13-15 U.S. Vicksburg, MS: Army Engineer Research and Development Center; 2013

[14] Downer CW, Skahill BE, Graulau-Santiago JA, Weston D, Pradhan NR, Byrd AR. Gridded Surface Subsurface Hydrologic Analysis Modeling for Analysis of Flood Design Features at the Picayune Strand Restoration Project, ERDC/CHL TR-15-X. Vicksburg, MS: U. S. Army Engineer Research and Development Center; 2015

[15] Massey TC, Pradhan NR, Byrd AR, Cresitello DE. USACE-ERDC coastal storm modelling systems in support of

hurricane sandy operations flood risk manage. Newsletter. 2013;**6**(4):2-3

[16] Ogden FL, Pradhan NR, Downer CW, Zahner JA. Relative importance of impervious area, drainage density, width function, and subsurface storm drainage on flood runoff from an urbanized catchment. Water Resources Research. 2011;**47**:W12503. DOI: 10.1029/2011WR010550

[17] Pradhan NR, Downer CW, Johnson BE. A physics based hydrologic modeling approach to simulate non-point source pollution for the purposes of calculating TMDLs and designing abatement measures. In: Leszczynski J, Shukla MK, editors. Practical Aspects of Computational Chemistry III, Chapter 9. New York: Springer; 2014. pp. 249-282

[18] Downer CW, Pradhan NR, Byrd A. Modeling Subsurface Storm and Tile Drain Systems in GSSHA with SUPERLINK, ERDC TR-14-11 U.S. Vicksburg, MS: Army Engineer Research and Development Center; 2014

[19] Richards LA. Capillary conduction of liquids in porous mediums. Physics. 1931;**1**:318-333

[20] Downer CW. Identification and modeling of important stream flow producing processes in watersheds [PhD dissertation]. Storrs, CT: University of Connecticut; 2002

[21] Downer CW, Ogden FL. Appropriate vertical discretization of Richards' equation for two-dimensional watershed-scale modeling. Hydrological Processes. 2004;**18**:1-22. DOI: 10.1002/hyp.1306HYDROLOGICAL PROCESSES

[22] Haverkamp R, Vauclin M, Touma J, Wierenga PJ, Vachaud G. A comparison of numerical simulation models for one-dimensional infiltration. Soil Science Society of America Journal. 1977;**41**(2): 285

[23] Brooks RJ, Corey AT. Hydraulic Properties of Porous Media, Hydrol. Pap. 3. Fort Collins: Colorado State University; 1964

[24] Huston JL, Cass A. A retentivity function for use in soil-water simulation models. Journal of Soil Science. 1987;**38**: 105-113

[25] Pinder GF, Bredehoeft JD. Application of a digital computer for aquifer elevations. *Water Resources Research*. 1968;**4**:1069-1093

[26] Alexiades V, Solomon AD. Mathematical Modeling of Melting and Freezing Processes. Washington: Hemisphere; 1993. 325 p

[27] Verdi C. Numerical aspects of parabolic free boundary and hysteresis problems. In: Lecture Notes in Mathematics. New York: Springer-Verlag; 1994. pp. 213-284

[28] Sergueev D, Tipenko G, Romanovsky V, Romanovskii N. Mountain permafrost thickness evolution under the influence of long-term climate fluctuations (results from numerical simulation). In: 8th International Conference on Permafrost. Philadelphia, PA: Taylor and Francis; 2003. pp. 1017-1021

[29] Marchuk GI, Kuznetsov YA, Matsokin AM. Fictitious domain and domain decomposition methods. Soviet Journal of Numerical Analysis and Mathematical Modelling. 1986;**1**:1-86

[30] Lovell CW. Temperature effects on phase composition and strength of partially frozen soil. Highway Research Board Bulletin. 1957;**168**:74-95

[31] Nicolsky DJ, Romanovsky VE, Tipenko GS. Using in-situ temperature measurements to estimate saturated soil thermal properties by solving a sequence of optimization problems. The Cryosphere. 2007;**1**:41-58. DOI: 10.5194/tc-1-41-2007

[32] Schulla J. Model Description WaSiM; 2012

[33] Zhang Y, Carey SK, Quinton WL, Janowicz JR, Pomeroy JW, Flerhinger GN. Comparison of algorithms and parameterisations for infiltration into organic-covered permafrost soils. Hydrology and Earth System Sciences. 2010;**14**:729-750. DOI: 10.5194/hess-14-729-2010

[34] Bolton WR, Hinzman LD, Yoshikawa K. Stream flow studies in a watershed underlain by discontinuous permafrost. In: Kane DL, editor. Water Resources in Extreme Environments. American Water Resources Association Proceedings; Anchorage, AK. 2000. pp. 31-36

[35] Rieger S, Furbush CE, Schoephorster D, Sumnmerfield H, Geiger LC. Soils of the Caribou-Poker Creeks Research Watershed, Alaska. CRREL Technical Report 236, AD 744451; 1972

[36] Viereck LA, Dyrness CT. A preliminary classification systems for vegetation of Alaska. In: USDA Pacific Northwest Forest and Range Experiment Station, General Technical Report PNW-106. 1980

[37] U.S. Department of Commerce. Monthly Normals of Temperature, Precipitation, and Heating and Cooling Degree-Days, 1941–1970. Asheville, North Carolina: National Oceanic and Atmospheric Adminstration; 1973

[38] U.S. Department of Commerce. Local Climatological Data Annual Summary with Comparative Data, Fairbanks, Alaska. Asheville, North Carolina: National Oceanic and Atmospheric Adminstration; 1980

[39] Monteith J. Evaporation and environment. In: Symposia of the Society for Experimental Biology, 4; Cambridge, England. 1965

[40] Monteith J. Evaporation and surface temperature. Quarterly Journal of the Royal Meteorological Society. 1981;**107**: 1-27

[41] Chapin FS, Ruess RW. Caribou-poker creeks research watershed. In: Chapin FS, Ruess RW, editors. Caribou-Poker Creeks Research Watershed: Hourly Soil Moisture at Varying Depths from 2000 to Present. Environmental Data Initiative. 2018. DOI: 10.6073/pasta/5207a6643ba e4f9792aa21180bc1c08d

[42] Farouki OT. The thermal properties of soils in cold regions. Cold Regions Science and Technology. 1981;**5**:67-75. DOI: 10.1016/0165-232X(81)90041-0

[43] Konovalov AA, Roman LT. Soil Mechanics and Foundation Engineering. 1973;**10**:179. DOI: 10.1007/BF01706681

[44] Engman ET. Roughness coefficients for routing surface runoff. Journal of Irrigation and Drainage Engineering. 1986;**112**:39-53

[45] Senarath SUS, Ogden FL, Downer CW, Sharif HO. On the calibration and verification of two-dimensional, distributed, Hortonian, continuous watershed models. *Water Resources Research*. 2000;**36**(6):1495-1510

[46] Rawls WJ, Brakensiek D. Prediction of soil water properties for hydrologic modelling. In: Jones EB, Ward TJ, editors. Watershed Management in the Eighties. Proc. of Symp. sponsored by Comm. on Watershed Management, I & D Division, ASCE, ASCE Convention, Denver, CO, April 30-May 1; 1985. pp. 293-299

[47] Eagleson PS. Dynamic Hydrology. New York, NY: McGraw Hill; 1970. p. 461

[48] Sutton OG. Micrometeorology. New York: McGraw Hill; 1953. p. 68

[49] Interagency Climate Adaptation Team. Adapting to Climate Change in Minnesota, Minnesota Pollution Control Agency, Document Number: p-gen4-0; 2017